STORIES NEVER TOLD

STORIES NEVER TOLD

A Country Girl, A City Boy
and Death in Vietnam

BONNIE COTTON

Cover Design by Peter Seglin

This book is a memoir. Everything happened as I remember. Any mistakes are mine. I have changed some names to protect privacy. All pictures are from my personal collection.

ISBN (print) 979-8-9903655-0-6

ISBN (ebook) 979-8-9903655-1-3

"Whoever destroys a single life destroys the entire world;
whoever saves a single life saves the world entire."

THE TALMUD

FOREWORD

I SMILED AND shed more than a few tears when reading *Stories Never Told*.

In January 2019, I had been a professional soldier for twenty-seven years. On active duty, living in Pennsylvania due to my duty assignment, I participated in planning the community's events surrounding Memorial Days. Each focused on one of Springfield's fallen sons. The event of 2019 was to be about Captain William Gary Chandler—a man who, at the time, I knew only from his high school yearbook portrait, a picture taken in Vietnam, and a short paper researched and written by a couple of high school students more than a decade earlier.

When Bonnie and I spoke, Bonnie was guarded. Beyond me asking personal and what I knew to be sensitive questions, I could tell she was concerned about the nature of the remembrance our community intended. Did we seek information to glorify his being killed in Vietnam at the age of 24? Until Memorial Day came, I didn't think Bonnie fully believed we intended only to celebrate his life and legacy rather than try to show that his death had some greater purpose.

I anticipated neither the deep emotional connection I'd make with the life of a man killed when I, nine months younger than their eldest daughter, was only two years old nor the significance to Bonnie and her family of the upcoming events. Over the next few months, Bonnie introduced me to her family and Bill's sister, Holly. Bonnie told me about Bill and their life together. I didn't know then how much of that part of her life, and her memories of Bill, she reserved only for herself. For nearly five decades, she coped with her grief through silence and by moving on. That changed in 2019.

In this memoir, with full transparency, Bonnie opens herself to the reader and speaks frankly of the less than four and half years she and Bill had together. She illustrates both the deep love they shared, the day-to-day challenges they, as a young Army family, experienced in the shadow of war, and the aftermath when that war took Bill and shattered her life. Through this story, we share Bonnie's grief. In 1972, she closed the curtain of her life with Bill to the outside world and their two daughters. We see why and watch her open the same curtain forty-seven years later.

Stories Never Told is not just about growing up, in-law relationships, experiencing love, tragedy, resilience, and healing. It's a memoir of social and historical significance. In addition to chronicling one man's experience as an Army officer in the late 1960s and early 1970s, his being wounded during his first tour as an infantry platoon leader, his service as a military advisor during his second tour, and the details of the fateful ambush on August 11, 1972, it describes the emotional cost of war and the impact his death had on those he loved and who loved him. Although written for a specific and personal audience, the rest of us are all the better for reading Bonnie's *Stories Never Told*.

RICHARD DEBANY
Colonel, U.S. Army (retired)
April 9, 2024

INTRODUCTION

William Gary Chandler
November 18, 1947–August 11, 1972

TAKING PHOTOGRAPHS, CAPTURING the sights around him, chronicling our daughter's growth, mastering the skills of developing, printing, and then mounting his black-and-white studies... all these things expressed who Bill Chandler was and how he saw the world. He dreamed and talked about leaving the Army and becoming a professional photographer. I planned to publish a collection of his pictures in the early years of widowhood. I couldn't think of a better way to show the world the gifts of the man I loved. However, without having him around to describe the images, I was never sure of their significance nor if I could complete the story they told.

As the years passed, I realized that a better way to tell Bill's story was in his own words. I have over one hundred letters that he wrote home to me. I have cassette recordings of his voice, including occasional mortar fire in the background. Thanks to my mother, I also have every letter I wrote home between 1966 and 1972, some with foreign postage from ten European countries, some with the return address of my college dorm, and many more sent from Pennsylvania, Baltimore, Washington DC, and Fort Lewis, Washington.

Over the years I have re-read Bill's letters from both tours in Vietnam and when he was training at Fort Bragg. I have listened to the audiotapes shared between us during his final tour. Each time, I re-lived moments of happiness, visited vivid images of where we lived and things we enjoyed together, and reclaimed a picture of our hopes and dreams. I ached for his love. I recovered the sense of honor and responsibility so strong I knew we would have made our life together work. I managed my anguish by tucking it away, seldom speaking of my grief.

Although nothing has been more life-changing than being widowed at twenty-four and finding my way afterward, I did not want to be defined by Bill's death, and I wanted to protect my daughters from the pain I felt.

On Memorial Day weekend in 2019, nearly forty-seven years after his death, American Legion Post 277 in Springfield, Pennsylvania, gave Bill the honor he deserved. Through the rituals of laying wreaths, the playing of "Taps," prayers of thanksgiving, words of remembrance, a parade, and a narrative of his life, I realized everyone's story needs to be told, retold, and spoken aloud. Time has given me perspective. The remembrances of Bill's Army comrades confirmed truths I knew but forgot. Memories lost in the shadow of grief have been recovered.

Our two daughters, Samantha and Abigail, and our two grandsons, Tristan and William, deserve to hear the stories of our great love and the life we planned. I pray the stories I have shared help them gain insight into the man I loved—who loved life—who loved them.

Grief never ends, and losses are never recovered, though they shift form while forming life. I know firsthand the human spirit is resilient. The following pages are the stories I neglected to tell, the memories of life together that only one other person, who is now long dead, shared with me. These stories are particular to my experience but, unfortunately, are not unique, for widows, parents, and children continue to grieve because of the violence in our schools and streets and the foolishness of war. May those who read, whether family, friends, or strangers, take comfort.

TABLE OF
CONTENTS

CHAPTER 1
RED LUGGAGE

1966

I ALWAYS KNEW I would leave the farm. I admired my mother and her sisters, farmer's wives who spent endless hours in the kitchen, from early morning bacon and egg breakfasts to keeping dinner warm while their men made one more round seeding a field. Watching them, I learned how to garden and feed chickens, cook and wash dishes, do laundry, and hang clothes out to dry on the line, giving them the fresh air of sunshine. I grew up ironing, sewing my clothes, mending whenever necessary, washing dishes, and dusting, constantly dusting. I knew that money was tight in years when the crop was poor or the price of grain dropped.

When I was in high school, the last one in the nest, my mother enrolled in college, wanting to be among the books she loved, seek a degree, and become a librarian. My father said, "I didn't marry you so you could go to work. Your job is to be at home, care for the family, and let me provide for you." After one quarter, she stayed home. I promised myself I would not have to choose between homemaking and working in the world.

The nice boys I grew up with might leave for college or join the service, but they were likely to return to the farm, the family land, expecting their wives to handle finances, keep house, fix their meals, and raise their children. As a farmer's wife, I would be expected to volunteer at church and school. There might be time for bridge parties and opportunities to visit the city once a month. There would be people who had known my grandparents and reminded me of the family name, how I was to behave, how to keep the family honor, or how to redeem the family reputation. Although I dated some of them in high school, auditioning each as a potential mate, I wasn't interested in marrying those nice boys.

The tiny town of Edwall, only three miles from my home and an hour from Spokane, Washington, consisted of a hardware store, a post office, grain elevators, a chemical company, a mechanics garage, an elementary school with four classrooms and a gym, one church, one tavern, and the general store. O.D. Byram's General Merchandise always had neighbors around the wood stove, the cold case filled with soda pop, and licorice on the candy counter. If times were hard, you could run a tab for a month or a season. If a woman stopped buying Kotex sanitary napkins, Virginia, the proprietor, would soon be sharing word of that woman's pregnancy.

My parents grew up there and never moved away. My mother's father was the postmaster, and to provide additional income for his nine children, he sold produce from his Truck Garden. It was common during the Great Depression for small farmers to raise an abundance of vegetables for that purpose.

My paternal grandfather died in a farming accident when my dad was fourteen. His mother was unable to keep the family farm, which altered my dad's future. Soon after my parents married, Dad worked as a hired hand on Carl Devenish's land and later was given a lease when Mr. Devenish died. In 1948, the year I was born, he and Uncle Clem, my mother's brother-in-law, began to share the lease and farmed together for the next twenty-eight years until my father's death.

Mother and her sister, Aunt Ruth, alternated driving us to swim lessons, just as they traded weeks of cooking for the harvest crew. Ruth and Clem's daughters, Cathy, a year older than me, and Marla, a year younger than me, were my best friends. In high school, we were cheerleaders together. We each planned to leave for college and not look back.

For high school graduation, my set of matching luggage wasn't the blue or white, hard-sided Samsonite my older sisters had chosen. Mine had to be red. It had soft sides. It could expand. It could hold more than I carried. It could include a future I had not yet imagined.

The summer of 1966 gave me five weeks to experience independence. I took my largest red suitcase to tour ten European countries in the company of a dozen other church kids, a pastor, and two chaperones. My mother's letters awaited me at every destination, but never the ones I wanted from John, my high school sweetheart.

Waking up in Switzerland on my eighteenth birthday, my traveling companions, who had gathered a floral bouquet from the hotel gardens, eased my longing for family. Later that day, I had a Sound of Music moment, turning circles on the hill above our tour bus while a flat tire was being changed. That night, I went to sleep in Austria.

In Italy, I discovered that the positive attention from handsome men soon became uncomfortable. I devoured pasta as the starter course, followed by meat, salad, and dessert in the cafes of Roma, Firenze, and Venezia, and knew I always wanted to eat that way. I shopped in the stores in Paris and viewed a more expansive world from the top of the Eiffel Tower, the steps of the Cathedral of Notre Dame, and the hills of Montmartre. I attended theatre in London. I grieved the devastation of war in the open shell of Coventry Cathedral, marveling at a statue revealing symbols of reconciliation and healing. I wandered the halls of ancient castles, imagining myself a lady in waiting or a prisoner in the tower. I gazed at the fields of the English countryside, connecting to my great-great grandparents' generations. My world expanded, filled with history, and my curiosity came alive.

I learned to drink strong coffee with three teaspoons of sugar to stay awake on the flight home. I made it through Customs with a ruby ring for my mother and a Swiss watch for my father. An airline strike canceled my original reservation home, and I managed to purchase a ticket from Seattle to Spokane, the airport nearest to the farm. I felt like an adult.

Returning home in the middle of the grain harvest, the beauty of rose and orange clouds at sunset magnified by dust in the air filled my heart, but the wonder of another, older world filled my head.

It was soon the day I would leave for college. My red luggage, plastered with my exotic hotel stickers as evidence of my adventure, filled the trunk of my parents' car. I was surprised by the tears that started as the dust rose in the driveway, the tires rumbled over the bridge, and we turned onto the county road.

I had chosen the University of Puget Sound (UPS) in Tacoma, three hundred miles away. It was a small, church-related school of two thousand students. I had been on the campus for church conferences. I also visited my sister Susan when she was a student. I loved the campus. It wasn't the state college for agricultural studies filled with would-be farmers. It was a significant enough distance that no one recognized the name of my hometown. No one knew the stories of my siblings, parents, or grandparents.

The unbidden tears continued to fall. I silently lamented that I wouldn't know anyone and no one would know me. I would no longer have anyone to call to check on what to wear to the first dance. I would never date again. I was accepted into the Honors Program but wasn't sure I was smart enough. I kept repeating I would become a small fish in a big pond. I wouldn't be able to talk to my mother every day or stop by the cookie jar on the way into the house.

We stopped midday and midway for lunch. When I arrived, it was after dinner and dark. It felt like I was the last person to show up on campus. We located my room, with my name and three others on the door.

Before I could enter my room, a girl grabbed me, called me by name, and told me how glad she was we would be living next door to each other. I had no recognition of who she was. It was awkward. She reintroduced herself. We had met at church camp. A year later, she would go home with me to attend my cousin Kevin's wedding, and I would introduce her to my cousin Wes, and they would get married. But at that moment, I was embarrassed.

Three strangers welcomed me into my assigned room. One lower bunk remained. I began to unpack. I had my high school wardrobe with a change of outfit for every day, ensuring nothing would be repeated for two weeks. My extra shoes, my records and record player, the lamp, pillows, stuffed animals, hairdryer, and cosmetics overwhelmed the space I was given. After a quick inventory, one suitcase was repacked to be sent home.

There were other "church" kids on campus whom I had known over the years, one high school classmate, and other kids from eastern Washington. I found a tennis partner who had a girlfriend still in high school. I learned to cook spaghetti on the hotplate in the laundry room, and Ruthi, my new best friend/roommate, and I had candlelight dinners on the ironing board, smoking Tiparillos. Making a long-distance collect call to "Edwall 613" was complicated and often confusing until an operator told me the three-digit number was a "ring-down" from Spokane. Being in the Honors Program meant I would get to attend theatre in Seattle.

I adjusted to living among evergreen trees, the vista of mountains, and months of rain. I yearned for home—a clear blue sky, green grass, quiet waters, spring floods, and the falling leaves in autumn.

Five blocks off campus, I found a store where licorice was on the counter. I walked through the neighborhood, wondering about the lives inside the lighted windows, missing a place to belong, and yearning for where and when I might start a family and build a home. I was unaware of the contradictions between my longing for independence and my desire to belong.

CHAPTER 2
EDUCATIONAL DNA

I DON'T RECALL the question of whether or not I would attend college ever coming up. It was a given, an expectation, my birthright, and a privilege. It may have been because women were finding their place in the world in the post-war, optimistic 1950s and 60s. Parents wanted and expected their children to have a more prosperous, easier life.

Yet, I also carried the legacy of pioneer stock, knowing my ancestors came from the British Isles, perhaps on the Mayflower. My maternal grandparents traveled in covered wagons as children. My grandfather, Arthur Edward Green, came from Indiana to Lincoln County, Washington Territory, in 1888. My grandmother, Clara Fern Walker, was born in South Dakota, came from Iowa, and arrived in Washington State in 1900. Both of them ended their formal education after eight or nine years of school.

Sadie and Flora Green, my grandfather's only sisters, became teachers. Sadie was thirty-two when she died from appendicitis. Flora, twenty-seven, died while caring for a family, including some of her pupils, during the flu pandemic of 1918. Neither of them had children and were sainted in my grandfather's memory.

Each of Ed and Fern Green's children, eight daughters and one son were expected to attend Teachers College, known at the time as Normal School. Two years beyond high school, one could become qualified to teach. Five of them completed their education and became teachers. The others attended business school before marrying or, in my mother's case, left at the beginning of her second year of Normal School to elope.

My father's educational options became limited when his father's legs were caught in farm machinery. His father died from gangrene and sepsis in 1929. From the age of fifteen, when they had to sell the family farm, Dad hired himself out to work on others' land, providing income where he could. His mother supported him and his two sisters by caring for other people's children and cleaning houses.

Wanting a trade, Dad enrolled in Diesel Mechanics School in Portland two years after finishing high school. Having written to him that she would die of a broken heart if he left without her, my mother convinced him to instead get married on the weekend he was to leave for school. Mother left a note for the sister she shared a room with to tell her she was borrowing her coat and eloping. After discovering he had left his wallet behind, Dad returned to Edwall to retrieve it, and Mother and her best friend went to an afternoon movie matinee to wait. That evening their wedding was held in the home of the local Presbyterian College president, for whom Dad's widowed mother worked. Grandma Monk and Kathryn, Mother's best friend, were the only ones in attendance.

It was 1934, in the midst of the Depression. Replacing two bald tires on his car depleted his savings, and Dad and Mother spent the first six months of their marriage sharing a shack with Rhoda and Herbert, another sister and brother-in-law, in Copalis Beach on the Washington coast. Both men were lucky to find a job cutting pulp wood for one dollar a day.

As part of the post-WWII "baby boom," I followed the path laid out by earlier generations' hard knocks.

Growing up, I struggled to answer the well-meaning but unfortunate eternal question of "What-do-you-want-to-be-when-you-grow-up?" I wanted my mother's life—to marry a reliable, hardworking provider, to stay at home with my children, to have free time to read, to entertain the ladies at bridge club, to volunteer at church, and to become an excellent cook—but that was not the answer I thought people wanted to hear. I also did not know of Mom's unspoken envy of the college-educated women in the community who had met and married WWII veterans. Across bridge tables and during afternoon conversations over coffee, she listened to the frustrations of these newcomers to farm life. I only knew these women as my friends' younger, better-dressed mothers. When asked, my closest friend's response was clear. Merry-K was sure she would become a nurse. My hesitant answer was that I wanted to be a teacher.

My brother, Bill, ten years my senior, referred to his sisters as attending college for an "M.R.S." degree. It was an excellent place to look for a husband, as my older sisters proved. Sondra married at the end of her junior year and Susan married in the spring of her senior year, neither one finishing their degree. Being a church-going kid, I selected a church-related college founded and supported by the Methodist Church. Preacher's kids were given financial aid to go there, certainly providing a pool of suitable young men to date and fulfilling my mother's dream of me becoming a preacher's wife and my dream of having my mother's life.

CHAPTER 3
ARIZONA SUMMER
1967

FIFTY YEARS PASSED, and I found the receipt for my first semester of college: University of Puget Sound, Tacoma, Washington. Current Funds, Received of Bonnie J. Monk, 9-21-66, $834.50.

Seems like nothing today. In the fall of 1966, it covered tuition, room, and board and told of sacrifice and commitment on the part of my parents. They paid the bill. My summer trip of five weeks in Europe had a price tag of $1,295, and I knew that my mother had given it to me as a gift, money she had saved from several summers as the substitute rural mail carrier. Initially, we were to travel together to New York for shopping and theater, and we probably would have returned with money in our pockets. Instead, I had an adventure that opened my horizons in unexpected ways. I was grateful, and being the daughter of my fiscally conservative father, who fretted about wheat prices and only paid cash for cars, I carried a sense of both gratitude and responsibility. I wanted to contribute something, and I tried to avoid asking for spending money on such trivial things as licorice and record albums.

In the spring of 1967, Dick, my brother-in-law, came to see me at school while in town to visit his folks. He shared stories about UPS, his alma mater, and I told him I was looking for a summer job to contribute to my tuition and living expenses. He suggested I come to Arizona for the summer and live with him and my sister Susan. As an Air Force captain with a cock-sure superior opinion of himself, he promised he could get me a job cashiering at the Base Exchange. I saw another opportunity to broaden my exposure to the world, or at least a corner of the United States.

I flew to Phoenix, Arizona. The promised job was not waiting for me. I encountered one hundred degree days, got acquainted with my two nephews, David, age two, and Christopher, age four, and started a job search. Meanwhile, Friday evenings began with happy hour at Luke Air Force Base Officer's Club, where the steak was cheap, the beer was cold and plentiful, and other young officers swarmed me like bees to honey. Susan, my introverted sister, drank because it lessened her inhibitions, and Dick was happier with her when she was relaxed, friendlier, and more talkative. In other words, inebriated.

My first experience consuming alcohol had been the previous New Year's Eve, drinking sloe gin and coke at a drive-in movie with my high school boyfriend, and that evening ended with me passing out in the front seat of his orange VW Bug while he changed a flat tire.

There was no sticky sweetness in the bitter taste of my first cold beer, but it was refreshing and cooled me off. It was also easier to flirt with strangers five or more years my senior. Miles away from home, my thoughts of John diminished once again. That first Friday night in Arizona, I accepted the invitation to a young officer's apartment with the promise of cold watermelon for dessert. I woke in his bed, no longer a virgin.

I was soon hired as a waitress, with no prior experience, at the same Officer's Club. It took me the rest of the summer to realize that after the second beer, my inhibitions disappeared. I would lose my sound judgment, and frequently, I passed out or threw up. The remainder of the summer went by in a blur. There was drinking while river rafting and being drunk at a party where I jumped fully clothed into an apartment complex swimming pool. I often needed to deflect sexual advances from the line cook while waitressing. With one month's salary and tips in my wallet, my financial resources had not grown much, but another set of life lessons influenced my life. I was no longer totally dependent on my parents' handouts. I believed I knew something about caring for myself. I was determined to find work at school.

We drove from Arizona to the family farm at the end of a very different summer. We stopped in Yellowstone National Park, where our cousin Cathy worked at West Thumb. She arranged lodging in a cabin in Grant's Village for Sue, Dick, and the boys but invited me to bunk with her and her coworkers.

The evening's entertainment was drinking beer while watching bears tip over garbage cans. When Cathy offered me my first cigarette, I accepted. Sue and Dick were both smokers, and I figured I might as well add one more vice to my summer misadventure. I am sure my choking cough revealed my foolish facade, but when I arrived home, I decided I needed to perfect my smoking skills before school started.

On my first night home, when I couldn't reach John and left messages with his younger brother, I went with other high school classmates to State Line, Idaho, where beer could be purchased at 18. As I popped off the top of my first beer of the night, Gina, the driver's younger sister, turned to me and said, "I always thought you were such a prude."

I replied, "I always was, my dear, always was."

CHAPTER 4
THE HOOK
MARCH 1968

I RETURNED TO college, another layer of naïveté removed, and began my sophomore year. The past summer in Arizona convinced me I had outgrown, and even betrayed, my boyfriend.

Unattached, I decided to pledge a sorority in order to have a social life. The rules for all initiates were the following: (1) They would attend study table for specific hours during the four months until they became full members, (2) They could only wear jeans or slacks on weekends or in the dorm, and (3) They would participate in all sorority chapter activities, including coffee "exchanges" and social events paired with designated fraternities, unless (or until) they were pinned or announced their engagement, in other words, "off the market."

My brother's assumptions were proving true. The atmosphere of finding a suitable mate was woven into the fabric of sorority life.

I also had more incentive to earn my own money. I had not shared my summer exploits with my parents and hardly felt I could ask for money to buy cigarettes. I found a job on campus, working ten hours a week as a PBX switchboard operator, answering incoming calls and connecting callers with numbered extensions.

Another new skill I needed to acquire while living in Tacoma's "big city" (population: approximately 150,000) was riding a city bus. I had no car. On the farm, my parents, my aunt, and my friends' parents provided transportation for social events and to the county swimming pool. Otherwise, I rode the school bus.

Once I got my driver's license, I borrowed my parent's car, especially on mornings when I missed the bus. In high school, there was always a boyfriend to drive wherever we wanted to go. In college, I needed to transfer from the bus that went downtown Tacoma to another bus that took me to the Tacoma Mall for shopping. It was a five-mile trip that took nearly an hour to complete. It added to my sense of being self-reliant, capable, and independent.

On the first Saturday in March 1968, I started the day with plans to go shopping for Rod McKuen's latest LP release. He had recently performed on campus. I had books of his poetry, including *Stanyon Street and Other Sorrows*, *Listen to the Warm*, and some of his previous recordings. Still, I needed the three-set compilation of *The Sea / The Earth / The Sky*, accompanied by The San Sebastian Strings. McKuen's words expressed the yearnings of lovers, the poignancy of loss, and all the fantasies of unrequited love. My morning trip was successful, and I debated between finding a sorority sister who wanted to go for a walk, taking in the campus movie, or facing an early evening of weekend homework.

My new Rod McKuen album was playing on the stereo when the phone rang. It could have been for any of us who shared the four-person suite, but my friend Alicia, who was working the switchboard, wanted to talk to me. She had a young Army officer on the line looking for a date, and there wasn't much more she could tell me about him. She knew I had decided my future would not be with my high school boyfriend, who was dropping out of medical school to join the Army. Maybe I could take the call. Nothing else to do, I thought. It might be interesting. Why not?

His name was Bill. He was from Pennsylvania. He was a newly commissioned second lieutenant. He told me he had already called the college across town but didn't get past the switchboard operator. He and three buddies were looking for girls they could take to dinner. Maybe I could help him out? I liked the sound of his voice and his laugh. I wasn't sure, but perhaps. However, times were different then, and I was oblivious to the stories of an as-yet-unidentified rapist and murderer finding his victims on college campuses in Seattle, thirty miles away.

Bill began to tell me about himself. His voice was gentle, the eastern Pennsylvania accent intriguing, his story irresistible, his candor opening my heart, his humor making me laugh. His mother was widowed and ran a public relations and advertising business to support Bill, his younger sister, and his brother. He had an older sister who was married and lived in Louisiana with her husband and one son. He had a German Shepherd named Dama, who was home in Springfield, a suburb of Philadelphia.

While I was finishing high school and traveling throughout Europe,

Bill had spent a year in the Merchant Marines earning money for college. We started college at the same time, but he left in January 1967 to enlist in the Army because he had a low draft number and thought he would fare better as a volunteer. He had his basic training at Fort Bragg, North Carolina, and Advanced Infantry Training at Fort Ord, California. He told me about his fascination with the Winchester Mystery House in San Jose, California. While I was in Arizona, Bill was in Officer Candidate School at Fort Benning, Georgia. In January 1968, he arrived at Fort Lewis, knowing he was on the fast track to lead a platoon into the rice paddies of Vietnam.

Was I attracted to the idea of a war hero? My brother served four years in the Navy, but Dad had been able to stay home as the only son of his widowed mother during WWII and was essential to the war work through his farming. Military service seemed inevitable in the Sixties—my generation was prime for the military—but I knew little about the reality of the conflict.

Was I intrigued by the nerve it took to "cold call" a women's dorm looking for a blind date? Was it his initiative that impressed me? Was it how he talked fondly of his mother and called her (as I called mine) every Sunday? Or was it, more likely, when he asked me if I knew of, or liked, Rod McKuen? I began to talk about my trip to the mall that morning, how McKuen had just performed on campus the week before, and which of his poems I liked best.

Then Bill recited,

"You have to make the good times yourself
take the little times and make them into big times
and save the times that are all right
for the ones that aren't so good."

I was hooked.

I asked him to call back in an hour, and I would let him know whether or not I found dates for him and his friends. None of my suitemates were around, so I went throughout the sorority from room to room and found Peg, who frequently hitchhiked to and from Portland; Barb, who captained the intra-mural soccer team; and Marilyn, whose math scores were at the top of the class. As usual, all were dateless on a Saturday night, willing to take a chance, have some laughs, and see what might happen.

The restaurant they took us to was unremarkable, the meal was forgettable, and since we were all under twenty-one, no alcohol was involved. I have no recollection of the names of the other three men, and I have lost track of my sorority sisters.

After our quadruple dinner date, Bill and I were the only couple who coupled. I imagine Peg, Barb, and Marilyn have the vaguest of memories, or they laugh at the preposterous notion that they ever did such a thing, warning their daughters that they were once quite foolish and naive.

Bill and I went alone in his 1965 black GTO, and he brought me back to the dorm only moments before curfew. His six-foot-two-inch height, large hazel eyes, megawatt smile, and dimples that defined his cheeks further hooked me. He was attentive and polite, laughed easily, and treated me with respect. We extolled the virtues of our mothers, talked about our siblings, shared our goals for the future, and fell in love in a few short hours.

Or, with the patina of time, it seems to have happened that way. We were soon spending an hour or more on the phone every night in the week, making plans to see each other again. When I went on a planned trip to Lincoln City, Oregon, the following weekend, I talked about nothing but Bill Chandler.

The Country Girl on the farm.
April 1968
through the lens of Bill Chandler's camera

The City Boy I took home to meet my family

CHAPTER 5
MOTHER'S LOVE

E VERY WEDNESDAY, AS I walked across campus to the Student Union Building to lunch, I often glanced at three lone figures standing in the circular driveway beside the fountain on the Jones Hall quad. On days it was raining, they held umbrellas. A sign was always in front of them, "Silent Vigil for Peace." It had been the winter of the "Tet Offensive," and anti-war protests were spreading across the country.

I recognized the three as classmates, pre-ministerial students, precisely the type of boys my mother wanted me to marry. I may have admired their courage or thought they should be embarrassed, but their actions did not intersect with mine, as I had my own soldier to worry about. Bill knew his orders to Vietnam would be coming soon. All the more reason to hold on to love and our committed relationship.

By the end of March, we knew much more about one another. Our childhoods were a continent apart, but our comfortable middle-class upbringing was similar—safe and carefree. He lived in a Philadelphia suburb, in a neighborhood of colonial-style brick homes filled with postwar veterans and their families. He told me stories about riding his bicycle behind the mosquito sprayer truck and running through clouds of DDT. I informed him about my summers catching frogs in the creek in front of our house and harvesting cattails growing along the banks. He played baseball every possible night after school while I rode the bus home, wrote stories, cut out paper dolls, or made families from the clothes models in Sears and Roebucks catalogs. We both had bikes to ride.

In high school, he was in the rifle club—I was a cheerleader and played tennis. His first job was as a lifeguard in the country club swimming pool. My first job was as a mother's helper, working on a neighbor's farm, cleaning the house, babysitting, and washing dishes. We both went to the Methodist Church with our families, a fact I shared in bold print as a postscript to my first letter home, introducing Bill to my mother.

We began to talk about getting married. Since I was headed home for spring break, I wanted Bill to meet my family. In my Sunday phone call, I mentioned Bill would bring me home but would only stay for the weekend, and we could visit Grandma and Grandpa Green.

Mother was nervous about making a good impression. She rightly interpreted my reason for introducing Bill to all the important people in my life. She ensured the house was spotless, and her usual perfect pies were cooling on the counter when we arrived.

In the two days we had in eastern Washington, we took the Sunday trip to Mill Canyon to have dinner with my grandparents. Bill returned to Fort Lewis, and I had the remainder of a week to extoll his virtues and discuss wedding plans. I wanted a lawn wedding on the farm, which meant Mother would be planting flowers, timing them to bloom by July 20, my chosen date. There were no spoken objections, though indeed, my mother's letter to my sisters says it all.

April 12, 1968

I'm excited about the prospect of having a wedding this summer and am anxious to get going on plans—that is, actually doing something rather than just thinking about it. I can't do anything until the engagement is announced and a definite date is set. Even though Bonnie prepared us with her enthusiasm about Bill, I'll have to admit that their announcement pretty much shook us that they wanted to get married so quickly. But after we got over the initial shock, we found we could accept the idea, and we were naturally very happy for Bonnie. We were favorably impressed with Bill. He seems to be a very responsible young man with definite life goals. We also think he shows good judgment and excellent taste by falling in love with our Bonnie and wanting to marry her... we do think he's lucky to have Bonnie fall in love with him. If you could see Bonnie, you would never doubt that she is in love. She is positively radiant. She has always been a happy-looking girl, but I've never seen her look this happy. Just hope and pray nothing will happen to destroy that happiness.

Though I've always felt very strongly that people should get to know each other thoroughly before getting married, I can also understand why young people hurry in these uncertain times. Although Bill is in the

Army and perfectly willing to serve his country in any way he is asked to, I pray he will not be sent to Vietnam. However, we cannot spend our time worrying about what will happen in the future. Right now, I feel we should do our utmost to help them have a lovely wedding and a good start together regardless of the future.

I planned to share the news with my sorority sisters when I returned to campus. I knew the ritual well. I selected the poem that would be read. The candle arrangement holding my engagement ring would be passed around the circle until it arrived in my hands. I would blow out the flame. I imagined the squeals of surprise and excitement, the hugs and tears. I would have accomplished the goal of every co-ed in the 1960s, my status would change, and I would be engaged.

Bill called his mother, Jayne, in Pennsylvania, asking for the family half-carat diamond and sapphire ring. Bill shared with me an abbreviated version of her concerns that maybe we were rushing this. If he went to Vietnam, I would be on my own, and indeed we could wait. A year wouldn't be too long, she said. Thankfully, Bill's visit with my family gave him enough background to tell her what kind of people I came from. It convinced her he was in love and that we wanted to be married. The ring was sent.

With plans to be married, it wasn't long before I was persuaded by his promise that he wouldn't ask me to have sex until I wanted to. Yes, I wanted to. Neither was the other's first, but what we experienced was new. Mainly because this time, every time, I was sober.

CHAPTER 6
PLANS CHANGE

W E WERE TOGETHER whenever possible that spring. Between Bill's work schedule—whether training enlisted men in counterinsurgency/counter-guerrilla warfare or on administrative duty, and my efforts at studying, trying to end the year with a presentable grade point average—we savored our time. With limited funds, we went to an occasional movie. We took day trips exploring the sights of the Northwest, Bill's camera in hand to capture Mount Rainier, Point Defiance Zoo, or the Pacific Ocean.

His first trip to the West Coast was near Fort Ord, California, while in Advanced Infantry Training, so the Northwest was all new territory for Bill. The Cascade Mountains and evergreen forests were a change from the deciduous trees and big cities of the Northeast. He declared the Atlantic Ocean superior to the Pacific, which I had yet to see. Most of our meals were provided—he ate all week in the Army mess hall, and I ate in the Student Union Building. On Sunday evenings, we often ate in his apartment. His specialty was fried bologna sandwiches. I never failed to return to the dorm before curfew.

Before I could announce my engagement, Bill had a change of mind. He began to think that since he had barely enough money to feed himself, he might be unable to provide for me. When we first met, he talked about Vietnam. With bravado, he was proud he was trained for the job and knew he would return home. The facts in front of us were that he had an equal chance of not coming back—through someone else's mistake or his own. Deep down, he was scared. He began to talk about not planning to be married because he wanted me to have a life of my own to rely on if he didn't make it home.

We came to a new understanding of our plans. My infatuation with being in love and my innocence about the world left no room for considering the possibility he might not return. I could not consider this a break-up, only a delay. Our parents supported what seemed to them a sensible decision while also sharing our heartache.

In the isolation of campus life, many of my fellow students were removed from the reality of war and, therefore, unconcerned. We were a generation both actively involved and equally oblivious. I began to see the solitary protestors on campus differently, feeling they were wasting time and not saving anyone's life. I put my trust in peace talks, believing they would miraculously find a solution. We began to refer to the Vietnam "conflict" as "that damn war."

Our new plan meant Janet and I would rent an apartment for the summer, still close to Fort Lewis, and Bill and I could see each other as long as he was stationed there. Janet, my suitemate and sorority sister, began looking for an apartment she and I could share in Olympia. She found a new, fully furnished, one-bedroom apartment for $115 a month. She would work for her dad, an Olympia doctor, and I began job hunting. I put twenty-five dollars, my share of the deposit, on the apartment and wrote home for an additional twenty-five dollars to reserve my room on campus for the following fall. I planned to be on campus for the first semester to participate in the life of my sorority.

I expected Bill and I would set the date to marry as soon as we knew when Bill would return from Vietnam. Then, I could pass the candle announcing my engagement. Being in school would get me one step closer to "a life of my own to rely on." Having a year to plan a wedding would keep me occupied and my mind off any concerns I had about "that damn war."

I no longer felt the agenda of finding a life partner. I assumed I had. I needed an education that would prepare me for work anywhere. I was ready to embrace a future that involved packing up my red luggage, following my husband's career, starting a family, and waiting for my happily ever after ending.

CHAPTER 7
HORMONES DON'T LIE

I T IS SUNDAY, May 19, 1968, two weeks before school is out.
I have a four-hour afternoon shift on the switchboard. Marty, my
suitemate and sorority sister, is bored and wants to come with me—no
harm in that. We can visit between calls. I expect it to be quiet.

A call comes in for another sorority sister. When I connect the call, I
tell Marty it sounds like Doug, Cheryl's boyfriend.

"Can we listen in?" she asks.

"I don't think that's a good idea," I reply. I am reminded of the party
line on the farm, knowing that our neighbor up the road was always apt to
pick up her phone when she heard the two long and one short ring of our
number, or anyone else's ring for that matter. We could always hear a faint
click and someone breathing.

"Please. I'll be quiet," she promises.

Against my better judgment, I help Marty put on the other head-
phones and open the line. The conversation is not as interesting as we had
hoped, and after a minute or two, I hear Cheryl say, "Do you hear that? It
sounds like someone else is on the line."

Quickly, I close the connection, and Marty and I laugh, but I feel
uneasy, fearful it might be reported, I might be reprimanded, and lose my
job. More than that, it feels like a betrayal of the sisterhood, and I am
ashamed and surprised that Marty is still laughing.

"I think you better go," I say. "You don't want to get in trouble." What
I am really thinking is, *I don't want you to get me in trouble.*

After my shift, back in my room, disappointed that Bill had not called
while I was on the switchboard and angry at Marty for causing trouble, I
settle in to finish an assignment for my French class. It is a class I struggled
with all semester, and I know my final grade is in jeopardy. I haven't con-
centrated much on my classwork meeting Bill in March, and finishing this
translation is hard work.

The hours pass as I struggle to concentrate. One by one my suitemates say goodnight and go off to bed. At midnight, I distract myself by glancing at the calendar and checking the schedule of final exams in the next week. I feel overwhelmed, uncertain of the outcome, concerned about my grade point average, and even more unsure of my future.

I replay the afternoon's events, then the last few months, and the tears begin to fall. My heart rate increases, and I cry harder but quietly, knowing I have no words to explain all that has happened. I do not want to confess my guilt over eavesdropping and don't want to call attention to myself. I am not sure I can stop the tears. I suddenly know what I need.

I make a plane reservation. It is five o'clock in the morning, but I call Mom to let her know I am coming home. I wake her up and give her the time of my arrival. All I tell her is that I am coming home. I pack a small suitcase. I call a taxi. I watch for it to drive up, praying that no one will see me leave, try to stop me, or ask me questions. I'm not sure I can speak without crying. The taxi takes me to the airport, and I board the plane.

Somehow, all of this happens without thinking much about it. I remember nothing specific, but I know it happened. It seems like a dream, a melodramatic, lousy dream. I wanted to be safe. I wanted unconditional love.

My parents meet my plane, embrace me, dry my tears, and take me home. Mom wants me to have breakfast, but I am not hungry. I sleep the remainder of the morning, and we see our family doctor in the afternoon. He is the doctor who had given me my childhood immunizations, stitched up the back of my head after I hit it on the steps at the county swimming pool, put thirteen stitches in my left knee after I overestimated my ability to leap over a concrete block during cheerleading practice, and set my broken wrist after I missed the last three steps into the basement while retrieving potatoes for Sunday dinner. His diagnosis is "exhaustion." He prescribes rest and recommends that I not go back to school.

The following weekend, Bill drives across the state. He is genuinely concerned and somewhat confused. It is good to see him, but I am cautious, still uncertain, maybe heartbroken, maybe something else. I know he cares, but does he love me?

For the next two weeks, I sleep a lot. I let my mother call the university dean of students and discuss how I might make up some of my classwork later and in which of my classes I would receive an incomplete rather than a failing grade. I spend time sewing, reading, and sleeping.

When I know classes are over and the dorms are nearly empty, my parents drive to Tacoma, help me pack the remainder of my personal items, and move me into the Olympia apartment. A month later, after being hired as a bookkeeper and salesgirl by a dress shop on Capitol Way, seeing Bill occasionally, and flirting with other men hanging around the apartment swimming pool, I seek the answer I already know.

On a Friday afternoon, I make the appointment, slip a birthstone ring onto my left hand to pretend I am engaged, take a bus to a doctor's office, and ask for a pregnancy test.

CHAPTER 8
GUESS WHAT?

L OOKING BACK ON this moment, I suddenly saw myself and my future differently. My embarrassment over my pregnancy may not have registered with this new doctor, but I left knowing I could not face him and would not be seeing him again. I even told him my (make-believe but wished-for) fiancé was in the military, and I would receive medical care once we married.

I knew I was not "one of those girls" promiscuous and careless enough to get pregnant. I realized how fortunate I was that my sexual encounters during my Arizona summer hadn't led to this. Abortion was neither legal nor readily available. Birth control pills were only for "those girls" who planned to have sex. I told myself that only "good girls" get pregnant unintentionally. I wasn't sure how I felt or what Bill would do. I certainly wasn't willing to talk about it yet. I invented a story for my roommate, telling her I had spent the afternoon at the Capitol building, seeing the crack in the dome from last year's earthquake. I wanted to talk to Bill but didn't feel it was the kind of news I could deliver over the phone, and he was seldom at his apartment to answer it anyway. We had plans for the weekend, and I had time to figure out how to tell him then.

It was a restless night. When and how I fell asleep, I do not know. I remember dreaming that I was in Arizona, living with my sister and her family, keeping the secret of my pregnancy from everyone so as not to embarrass my parents or bear the shame for myself. In my dream, I let my sister raise the baby, giving her the girl she wanted but did not have. It felt like my only solution.

Janet, my roommate, invited me to come with her Saturday morning to visit her dad and stepmother across town. I enjoyed their company and usually would jump at the chance for a home-cooked meal and the opportunity to sit on their deck overlooking Budd Inlet on the water. I assured her that Bill would show up sometime later and she should spend as much time as she wanted with her family.

I wish there was a dramatic or romantic moment to recall from his arrival. He did not sweep me into his arms or weep. He was neither angry nor surprised. He took responsibility for what we had done together, and we spent most of the day discussing wedding plans. While we were still talking, Janet returned.

Bill immediately announced, "We are getting married."

Any doubts I may have had about his delight in starting a family vanished. Our decisions were more practical than romantic. We chose our wedding date for the first Friday in August because it was the day after payday. We decided to get married at Fort Lewis since he wouldn't have the time off to travel across the state. His mother's diamond and sapphire ring was still in his possession, waiting to have the diamond set as a solitaire engagement ring with an accompanying silver wedding band. He planned to call his mother in the morning, and my parents were arriving the next day on their way to the Washington coast for a fishing trip. We did not rehearse how or what we would say, only that we had decided on the date and the place.

When Mother arrived in Olympia, I immediately shared our plans. Janet, my roommate, had agreed to be my maid of honor the night before, and I planned to ask Merry-K, my oldest childhood friend, to be my attendant. Parmalee, Merry-K's mother, offered to host the reception in the backyard of their home in Olympia.

Mother, Parmalee, and I discussed the details over my lunch break from work, or rather, I agreed to their plans. There was a stand-by possibility of holding the reception at the Governor's Mansion in case of rain since Merry-K's father was serving as head of the Department of Fisheries under Governor Dan Evans.

It was not the era of weddings being the extravagant production they have become. I had once harbored the fantasy of walking down the aisle of a cathedral with a fifteen-foot train of satin behind me while a chorus of nuns stood behind a screen, singing, "How do you Solve a problem... like Maria" ala *The Sound of Music*. Unlike a high school classmate, I had not spent my senior year planning my wedding to the football team captain, nor did I make my wedding dress in home economics class and model it at the spring fashion show. I may have once expected a church wedding and a reception in the basement, with my aunts in attendance to pour the punch and coffee, serve the cake, and pose in a line for pictures together. In April, my mother and I had begun to plan a wedding in the front yard of my childhood home. On that July day after telling Bill I was pregnant, my only desire was to be married.

I returned to work with only a few minutes to spare. We began to say our goodbyes on the sidewalk in front of the dress shop. My mother hugged me and said, "Parmalee mentioned that she hoped you had considered doing something about birth control."

Surprised, I thought, *Doesn't this sudden change of plans make it obvious?*

I smiled and said, "I think we have that covered."

Mother held me in a final hug, and I returned to work.

Why didn't I tell her? Why couldn't I say the words out loud? It didn't seem like I had time to explain everything. Yes, I thought, she might have guessed or figured it out. Why didn't she ask me what I meant? I knew I didn't want it to seem we weren't madly in love and that our marriage was simply due to a sense of obligation.

I remembered my sister's unplanned pregnancy and her rush to marry four years earlier. I witnessed the tears and experienced the disappointment expressed by both my parents. Sue was our father's favorite. They were more alike than he and I, both introverts. I would have said he always liked her best, and I wanted to protect myself from losing his affection, feeling we weren't that close anyway.

My mother must have made a phone call that night, for the announcement of our engagement and wedding date appeared Thursday in the weekly county paper. The information generated the organization of a bridal shower. The week of the bridal shower included our visit to the bridal department of the Crescent Department store in Spokane. By then, I was wearing a diamond Solitaire on my left hand and a radiant smile on my face.

After the bridal shower, I wrote my mother a letter, apologizing for not telling her sooner but letting her know I was expecting a baby in February. My flippant response to her suggestion about getting on the pill triggered what she called "a mother's intuition."

Our dance around the subject continued when she wrote to tell me she had decided not to mention my pregnancy to Dad until after the wedding. Her letter was also my birthday letter, with her wistful desire to impart words of wisdom, words I have cherished, words that I embraced and hopefully embodied in my own mothering:

July 27.1968

Dearest Bonnie,

Your letter did not come as a surprise to me. You "told" me on the street in Olympia when I suggested the use of the pill. I could have asked you the direct question but decided not to and was willing to accept your decision of whether to tell us before the wedding or just let us find out when the baby came.

I have long ago learned that nothing is "perfect" in this world—so we have to accept things as they are and get what joy we can from near perfection.

I've always felt you have understood my feelings towards you and the rest of my children more fully than any of them have, so I expect you know without my saying it that my love for you never changes, and your happiness is my happiness.

Just one more thing I want you to know. There are no tears! I think perhaps not even you fully understand my heartbreak over Sue's situation. Someday, I may try to explain it to you. Suffice it to say that this time there is no little sister to be hurt.

We want you to have a lovely wedding with all the excitement and joy that any bride has ever had.

Even though I have been accused of being an incurable romanticist, I do have intelligence enough to know there is no such thing as "happily ever after."

Perhaps all I should say is "God bless you," and continue my prayers that life will treat you kindly and that when trials and tribulations do come to you, you will be able to cope with them and emerge a stronger person, more capable of appreciating the good things that come your way.

But I cannot stop there, and I would like to say I have a great deal of faith in your capabilities of coping with life's problems and your common-sense realization that sometimes you can't solve things all by yourself and need to turn to others for help.

I hope you will always remember that you are an individual, a person in your own right, and though you may have faults and weaknesses, you are a person of worth. I don't mean you should be smug and self-satisfied; everyone has room for improvement. Just don't let anyone make you feel like you are nothing.

I think sometimes children tend to forget that their parents are people too and have experienced the same emotions and feelings, the same temptations, frustrations, and satisfactions that they are facing in life. Sometimes, we, too, have floundered, failed, and came through with flying colors. I guess the idea I'm trying to get across is that we have a fairly good idea of what life is like, and have more understanding and sympathy in our hearts than our children give us credit for.

This leads up to what I most want to say—that no matter what you do or have done, no matter where you are and what your circumstances are, there are two people who love you very much and will always be standing by to help you—as long as they are living.

We have never intended to make any of our children feel they had to be a certain kind of person to please us. The longer I live, the more I realize that each person is a unique individual, and each one must find his or her true self and live accordingly. My love and concern for my children will never cease, but I know they are separate persons and must live their lives in their own way. If I should make suggestions or give advice, it is done lovingly, and they have every right to accept or reject my advice as they see fit.

Much love, Mom

When she shared my letter of confession with Dad, he was not surprised but very understanding. They agreed they were glad their daughters were such loving persons, enjoyed sex, and couldn't say "No" to the men we decided we were in love with and wanted to marry. He told her, "Bonnie takes after her mother."

Years after my father died, Mother and I revisited my decision to try to minimize their disappointment. She told me she had not been as upset because "this time there was no little sister to be hurt."

Only then did I hear the story of my Aunt Rhoda, her older but closest sister, who became pregnant in high school in 1929. Rhoda was sent away to a home for unwed mothers, delivered her baby, and returned to finish high school. No one talked about it; everything was covered in secrecy and lies.

Only then, when she was in her seventies, did I hear the story of my mother's abortion in 1940. Working as a hired hand, Dad felt they could not afford to raise more than the two children they already had and persuaded her to terminate the pregnancy. She acquired a life-threatening infection and nearly died. Thankfully, she could have two more babies later, including me.

These untold stories were the fears on her mind and heart when her daughters became pregnant. Her heartache came from her memory and experience, influencing her desire to protect her children from pain.

CHAPTER 9

GOING TO THE CHAPEL

Our Wedding Day, August 2, 1968

W E HEADED TO the courthouse for our marriage license on July 26, 1968, the day after my birthday. We were both twenty years old. Bill would turn twenty-one in November. He needed permission to get married, and I did not.

Washington state law required men to be twenty-one and considered a woman old enough at eighteen for consent to marry. Bill had to ask his mother's permission, which came with her notarized signature that week. He had to request approval from his commanding officer, a requirement for receiving marital and financial benefits. It was a blow to his ego and a subtle reminder that we still had some growing up to do.

Jayne, my future mother-in-law, wrote an apologetic letter to my parents in mid-July, explaining why she would not be able to attend our wedding. She had just returned from a two-week vacation with Bill's younger brother and sister and didn't have excess funds for a trip out west. It was also a letter of introduction, giving enough of the family background story so that both she and my parents would be comfortable with our marriage. She wrote this about Bill:

July 18, 1968

Bill is the second oldest (of four children)—and has been a joy as a son. He has been a boy you could explain things to—and reason with and has been easy to raise and live with. He's been independent—earned his spending money and many of his clothes through high school—and went to sea to earn money for his college—even tho I pleaded with him to let me send him. He said I worked too hard and he wouldn't feel right taking the money. Of the four children, Bill has been the closest to me... and I have thoroughly enjoyed him. I am so glad he's found someone who loves him and whom he loves deeply... Bill has earned a long tether for some years now, but I guess it's his turn to be cut loose.

Even though she sent us the cost of her airfare as a wedding gift, she soon changed her plans and arrived Thursday afternoon, in time for the rehearsal that evening. Bill warned me that his mother was a "large" woman, yet I was still surprised when she stepped off the plane. Her weight was more than twice mine then, and I tipped the scale at 140 pounds. Her personality was equally abundant, embracing me and offering love and support for our marriage.

My other impressions were that Bill's brother, Christopher, at sixteen, was a gentle soul with a subdued version of Bill's smile, and Holly, age eleven, seemed reserved and shy in the shadow of more prominent figures, but pictures reveal the adoration she felt toward her big brother and wonderment over our wedding.

Sister Sue came from Arizona, and she and Mother shared a hotel room. My brother and father had time to drive across the state on Friday to arrive for the 7 p.m. ceremony and not miss more than one day of the wheat harvest. Bill spent the morning with his family, while Sue, Mom, and I went to the beauty salon for hairstyling appointments. I waited patiently for an hour before I spoke up, asking when I would see a stylist.

"We only have two appointments for you on the schedule," the receptionist said.

"But I'm the bride," I stammered, stalling my tears.

"Oh my, we are so sorry. Let me see what we can do...."

They worked me in, styling and lacquering my hair to stay in place through the remainder of my big day.

Just as my wedding dress was off the rack, Janet and Merry-K were dressed in matching readymade lavender dresses, with short-veiled headpieces, carrying lavender and white daisy bouquets. To Jayne's displeasure, Bill had ordered and presented me with a small bouquet of white roses and carnations. She had let him know which aspects of the wedding were the groom's responsibility according to Emily Post's book *Etiquette*. He was supposed to provide for the rehearsal dinner (which we skipped due to time schedules and cost), the rings, and my bouquet. She assumed I would tell him what I wanted, which I did not, and neither did I care. I was pleased with what he provided, which fit his budget. Remember, I wanted to BE married. How it happened did not matter.

While I was dressing, Bill was organizing eight fellow lieutenants for the arch of sabers, which would complete the military aspect of our ceremony. In dress blue uniforms, white gloves, and holding swords provided by the chapel, Bill instructed them on the commands to be given. Dad watched them practice, pleased to see Bill's leadership, impressed by his ability to take charge.

The ceremony was brief and basic, getting the job done. While we let the wedding party and our guests leave the chapel and descend the steps ahead of us, the saber-bearers took their place in pairs. Tom Slining, John Eversole, Michael Bacchilega, James Vojtecky, Steve Sutherland, Todd Ruedisili, John Zimmerman, and John Reed, all second lieutenants, pivoted to face each other at James Vojtecky's command, "Center, face." At his order, "Arch, sabers," each saber was raised with the right hand until it touched the saber's tip directly opposite.

Bill and I watched from the chapel doorway, then walked underneath the raised sabers together. As tradition requires, while we made our way to the end of the arch, the last two men lowered their swords to stop us while the saber bearer on the right gave me a gentle "swat" on the rump and declared, "Welcome to the Army."

Besides Bill impressing his father-in-law with his leadership, the moment thrilled his wife. My smile in the pictures reveals my delight.

Having changed the venue of my wedding to a chapel on a military base instead of the yard of my childhood home meant that the wedding was small, although five of my mother's seven sisters—Marian, Rhoda, Lois, Ruth, and Meryl; her brother Robert, and my father's sister, Dorothy, were faithfully in attendance. Only six of my thirty cousins, three of Bill's family, a few college friends, and a cadre of fellow soldiers and their dates celebrated with us.

My oldest sister, her husband, and their five kids missed the ceremony because of Seattle traffic, arriving late to Fort Lewis and unable to find the Division Chapel on Division Road. Lost from Fort Lewis to Olympia, they arrived at the tail end of our garden reception after I had tossed the bridal bouquet. Cousin Cathy, whose wedding was scheduled four weeks later, caught it.

I changed into a turquoise dress with a matching jacket for my "going away" outfit, but we went no further than a hotel in downtown Tacoma. It would be the first time we spent an entire night together, and I lay awake most of it, reliving the day, imagining all our wedding guests walking in on us in our hotel room.

CHAPTER 10
PLAYING HOUSE

B ILL QUALIFIED FOR base housing, after our marriage, but the orders
had not come through, so for the first week of our married life, we
had minimal furniture in an apartment complex still under construction.
With only one car, Bill went to work, and I spent the day alone, planning
dinner and writing thank-you notes at a borrowed card table.

I hadn't given notice at the dress shop. I simply did not show up for
work on Monday, forfeiting my last paycheck, yet blissfully happy to be
married.

By the middle of August, we moved into officers' quarters, with the
bland designation of Quarters 2506E, part of a four-plex. We rented a
bed, a kitchen table with two chairs, an orange sofa and chair, two lamps,
two end tables, and a coffee table. Our entertainment center was my college turntable, Bill's reel-to-reel tape player and speakers, and a tiny black
and white TV.

Soon, our first major purchase was a console stereo, complete with
monthly payments. My parents returned after harvest, delivering the
dresser and canopy bed with white eyelet canopy and bedspread, which
had been a gift for my sixteenth birthday. They brought a few late wedding
presents, completing our needs in the kitchen, including twelve settings
of Noritake China, white with uneven silver stripes on the rim.

Of course, we soon added Max, a German Shepherd puppy, to the
household, and besides waiting alone all day and planning dinner, I now
needed to walk the dog or chase it down when it got loose. I was only
familiar with farm dogs, who spent their time running free and were never
allowed in the house. Complaints by a neighbor to the military police
meant he became my responsibility to walk on a leash. He rode with me
in the car whenever I ran an errand.

Fortunately, Max loved car rides and would sleep while I was shopping
or going to an appointment. He behaved well the day I drove into Tacoma

and returned to the University of Puget Sound campus to visit with Janet and other sorority sisters.

I had hoped to make a surprise announcement of my marriage to replace the engagement announcement I never got to make. Word had spread, and I was greeted warmly with congratulations. Since I was only four months pregnant, there was no baby bump. I had not seen Janet in the six weeks since the wedding, and as she walked me to my car, she told me I looked so skinny she was afraid I had miscarried. I reassured her that she could still look forward to a February surprise, and then she could share the news with our sisters.

I was invited to join the Officer's Wives Club and eagerly attended teas and meetings, suddenly aware I was at least ten years younger than most of them and that the ranking of officers extended to their wives. In the evenings, Bill invited other soldiers over for lengthy cutthroat games of Monopoly. With a baby on the way, Bill began planning for life beyond military service and signed up for a correspondence photography course. Taking his 35 mm single-lens reflex camera with us on weekend excursions was natural. Fulfilling course assignments gave us destinations. Access to a dark room and developing equipment happily occupied his free time.

My childhood dream of being a housewife, caring for a husband and children, being provided for, and living in a supportive community was intact. Bill was showing himself to be a thoughtful provider, a responsible man who shared my desire to raise a family and create a happy home. When Bill announced the news of my pregnancy and our plans to marry to a friend, the response had been to tell Bill he knew how to take care of Bill's "problem." Our situation was unexpected, but Bill's immediate answer was that he was excited about becoming a father and looking forward to our life together. We didn't have a "problem" to be solved.

CHAPTER 11
WHY, OH, WHY, PENNSYLVANIA?

FIVE WEEKS AFTER the wedding, Bill received his orders to Vietnam. We considered renting an apartment near Jayne in Springfield or Philadelphia, where I could prepare a home for his return from Vietnam. He planned to leave the Army and use the GI Bill to attend college.

Once again, our effort to be adult and responsible and make our own decisions was altered, this time by Jayne's invitation to stay with her. The room at the top of the stairs, shared by Bill and his brother Chris, was vacant while Chris attended Milton Hershey School. Chris had written to his mother:

I've been thinking about how Bonnie will feel so far away from him at Christmas time... away from her family and with Bill gone. I hope you'll convince her not to rent an apartment and give her my room. I don't care where I stay when I am home, as long as I'm home. She'd be much happier with us, making Christmas cookies, trimming the tree and all.

We accepted her offer to stay in the bedroom, and Jayne had the room repapered, painted, and carpeted. There were twin beds and a small room attached, cleared out, and prepared as a nursery.

Bill had forty days of leave before reporting to a Jungle Operations Training Program at Fort Sherman in the Canal Zone, Panama. We prepared to move out on the first of October.

Having lived in military quarters for less than two months, I imagined it would be easy to return our rented furniture, let the movers pack up, and clean up after them. I'll admit my pride was injured when Bill informed me we needed to hire someone to clean our place well enough to pass a military inspection. When our household goods were packed, I left for the farm and had an extra week to visit my parents.

Max needed a home. My parents reluctantly agreed to keep him. The last dog that had been on the farm was my collie, Fabian (yes, named after a singer who was famous when I got him on my twelfth birthday). There were always feral cats in the barn as well as yard dogs on the farm. I grew up with a succession of red cocker spaniels, each named Turpey. My father couldn't see spending money on pets, so when Fabian got sick while I was touring Europe, there was no one to champion a visit to a vet, and I came home to the news that Fabian had died.

"Beastie," my sixteen-pound tabby cat, had been given the rare privilege of being a house cat, left outside when no one was home. Accepting a German shepherd puppy, with the expectation that he be a house dog, was a stretch for my folks. It was one of many ways everyone did whatever it took to send Bill off with the sense that he was supported and loved.

When Bill came to the farm from Fort Lewis, we had another week, including taking studio portraits to complete our wedding album, which had been only a collection of enlarged candid shots. On another day, we drove to Palouse Falls (carved more than thirteen thousand years ago on the Ice Age flood path) and Grand Coulee Dam. I wanted Bill to see Washington's self-proclaimed "eighth wonder of the world." Being early fall, both were less than wondrous in the volume of cascading water. Then we gave ourselves a week to drive across the country.

In truth, two stories explain my decision to stay with my mother-in-law while Bill was in Vietnam. The one shared most frequently was that I did so because Bill wanted me to, and I wanted to please and honor his desire. As we prepared to leave Washington, he needed reassurance that it was what I wanted to do, understanding how hard it would be to be away from my family. I reasoned that getting acquainted with my husband's family was essential and that being near them would help me remain near Bill. It would allow us to save money. And, as I told my parents, the girl with the red suitcase loved the adventure of travel.

The other story, my heart story, was the conflict of how I was known in my small world of growing up as a "church girl," with a reputation of being a bit of a prude, measured against the image of the "bad girl" who would engage in sex before marriage. I did not want to be the subject of gossip, feel the judgment, or cause embarrassment to my parents. Though my grandmother had sent one of her daughters away to a home for unwed mothers at the age of sixteen, and her youngest daughter went to Seattle to live with her big sister while she was pregnant at seventeen, she began to respond to her grandchildren's pregnancies with, "The first baby can come anytime; after that, they take nine months." I thought if I went away, no one would count out the lack of appropriate time before my baby arrived.

At twenty, determined to be independent, responsible, and able to make decisions, I was unaware of how it felt for a mother to watch her daughter take flight. As that year unfolded, my appreciation grew for the unconditional, nonjudgmental, unrestricted love my parents had given me. I expected no less from life. I tested and stretched the limits. The bond between us never broke but grew more substantial.

Heading east across the country was a carefree adventure. We drove on Interstate 90 through Washington, Idaho, and Montana, stopping at the Custer Battlefield, then detoured south to see Devils Tower in Wyoming. As I continue to do, I picked up brochures at every National Park and Monument, and I still have the ones from our "honeymoon" trip.

We stopped at Mt. Rushmore, which seemed to have decreased in size since I had seen it as a five-year-old. I was fascinated by the Badlands, having been interested in the life of prairie dogs since reading about them in the *World Book Encyclopedia*. We stopped at Wall's Drug and the Mitchell Corn Palace in South Dakota. We took our time on Interstate 80, or it seemed to take more time to drive through the open spaces of the Midwest before entering Minnesota and Wisconsin, into traffic, and through large cities. We stopped to visit two of Bill's uncles and their families in Grosse Pointe, Michigan, catching up with relatives he had not seen since his father's death nine years earlier.

We arrived at 46 Longview Drive, Springfield, Pennsylvania, Bill's childhood home. It was the week of Halloween, and we were caught up in visiting neighbors and Bill's high school buddies. Bill took me to the Reading Terminal Market in Philadelphia for my first Philly cheesesteak.

I soon learned I had to adjust my vocabulary, asking for "pop" in an area where carbonated beverages were known as "soda." My lifelong pronunciation of the Monopoly space as the "Reading Railroad" also needed to be corrected to reflect the origination of the game in Atlantic City, close to Reading (Red-ding), Pennsylvania. When we picked up a paycheck and shopped at the Post Exchange at Fort Dix, I added New Jersey to my growing list of places I now had been. We took comfort in being together yet imagined how our soon-to-be separation would be unbearable. We knew we had only a month before he left.

We were invited to dinner by parents of classmates and former teachers. One evening, Bill leaned toward me and said he wanted to tell his sixth-grade teacher that we were expecting a baby. I smiled at him and reassured him it was visible, and she already knew. I still regret not letting him announce our good news with the pride he felt.

On Bill's twenty-first birthday, I spent the afternoon making a three-layer heart-shaped devil's food cake. He returned from picking up his liquor card (a Pennsylvania proof of age thing) and shopping for something indispensable to buy. He came in empty-handed to find his loving, devoted wife in tears because the cake slid off the plate and broke into crumbs while I was frosting it. It hadn't helped my disappointment that Jayne pointed out my error in not trimming the layers evenly. I felt like an abject failure. Pregnancy hormones, fear, and despair did a number all at once. Writing it off as another "just-married" disaster took a while.

With only ten days before Bill was scheduled to leave, he stayed up late every night and had difficulty falling asleep. We didn't need to talk about the coming separation on our minds, and there was no way to be reassured that there was nothing to worry about. He worried about his capabilities as an officer in a new situation—much more than the actual fighting. He was a very proud man but not quite sure of himself. My tendency to worry about ifs, ands, ors, and maybes didn't help, but our conversations about "What-if-things-had-worked-out-differently" always ended with the agreement that we held no regrets about our marriage. Our motto was, "It makes life more pleasant to decide to be happy."

CHAPTER 12
PLEASE WRITE

MY MOTHER WAS a faithful correspondent. She could be counted on to write to each of her children weekly, and I did my best to answer her letters. The rhythm was part of my life. She saved my letters and her monthly contribution to the Round Robin that circulated between us—my mother, my two sisters, and me. I am grateful to have all of my mother's letters to refresh my memory and correct it as needed.

It was easy to promise Bill I would write to him every day. He said he wasn't much of a letter writer, but he would do his best. I sent him off with a portable cassette recorder and half a dozen tapes. If he couldn't write, I hoped to hear his voice and receive something once a week. I also promised to do my best to record my thoughts occasionally.

When he left Springfield the day after Thanksgiving, he was flown to Panama. He would spend two weeks acclimating to high temperatures and humidity and learning about warfare in the jungle. We had a tearful goodbye at the house, Bill telling me that he appreciated my being a "big girl" about all this.

A friend drove him to the airport, and I made myself busy, dreading facing my first night without him.

His flight took off at 3 p.m., and he had a stopover in Charleston, South Carolina, waiting for the next leg of his journey, which left at 3 a.m. I had just finished writing to my mother when Bill's surprise phone call lifted my spirits. He had run into three or four men with whom he had completed Officer's Candidate School at Fort Benning the previous January. As expected by each of them, orders came through for Vietnam on schedule. They were keeping each other company at the Officer's Club.

Two weeks of Jungle School were cut in half, and he returned home for a surprise bonus of another five days together. By then, the Christmas tree was up and decorated with gingerbread men, birds, Santa Clauses, and bows, a stark contrast to the old-fashioned tinsel-draped trees of my childhood, chopped down from my grandparents' land.

Bill departed on December 19, 1968 for his first tour in Vietnam.

I had no address, no place to send him mail. Still, I wrote a three-page letter the next day and then began to record my voice. I was self-conscious at first, then becoming more relaxed when I turned on the stereo and could listen to Barbra Streisand sing "People Who Need People," or one of our favorite love songs by Henry Mancini, or George Gershwin's "Rhapsody in Blue," or Rod McKuen's gravelly voice in spoken word or song. They muffled the sound of my voice in case anyone was outside my bedroom door, and I began to imagine us together, just talking.

Despite his protestations that he wasn't much of a writer, a steady stream of letters began to arrive. The first letter I received was an honest reflection of his fear and his apprehension.

Dearest Kitten

It is 19 Dec 1968, and the time in Calif. is 2130.

I pray that our first child is a girl. I don't think I could stand to see my son go off to war; not a war like this.

I spent about five hours at Travis Air Force Base sitting, thinking, and drinking coffee. For me, it wasn't too bad, but I pity the soul who would start on such a venture without knowing anyone or having a good friend to talk to. Not just anyone but someone you can talk to and know that he understands.

We're on the plane now, a Boeing 707. One hundred sixty-three strong. Army, Navy, Marines, Air Force. Not a unit here, just individuals, all with doubts, fears, and memories of what they left.

163 people going to a place to fight in a so-called war.

A war with no end in sight and no gain intended.

163 boys, the cream of the crop and the sons of parents who love them.

As I look around, there can be seen the entire group with their mixed emotions.

Those that are loud and raucous for one of two reasons. They don't know what they're going to or are so scared that silence would shatter their facade.

The quiet who talk of what they've heard about Viet Nam, about their wives, families, their past. And there are the silent who are scared, for they know what lies ahead, and talking to these people only frightens them more. I haven't classified myself yet.

163 leaving for Viet Nam for a one-year tour.

Those who are silent, quiet, and loud know that after one year, no one could assemble those 163 for a return flight.

They instead wonder how many killed, how many wounded, how many maimed, how many hurt so bad that they'll be vegetables the rest of their lives, and how many will live one more year.

I predict nothing but wonder much.

There are only three classes: those that make it, those that don't, those that make it back but not all in one piece. Apathy is a scarce commodity right now. They all ask, which class will I be in?

We have all of them here. Privates, NCO's and officers.

I wonder how the private feels. He probably wonders what he'll be doing and where and will his leader be competent.

The NCOs, I imagine, wonder the same thing, but with more responsibility added in.

The officers, I know how they feel. Some wonder about themselves. Do they know what to do? They'll be given men over there; will they finish with all that they started with? When everyone looks to him for the answer, the decision, the plan, will he have the right one, the one that will kill the fewest? So young, so new, so much in charge.

Now is the time for serious self-evaluation for many.

It's 2200 hrs now. I rest and think and wonder, about them, about the upcoming year, and about me.

Love, Bill

His letters would be written as much as a week before I received them. His postage was free, and his letters journeyed from his base camp in the jungles of the Mekong Delta through San Francisco before heading to the East Coast. He wrote details of life in a combat zone, descriptions of the country and people he encountered, reports on the weather, and frequently of his fears and feelings.

As he traveled closer and closer to his military compound, he remarked, *"Everything around here that has anything to do with the military has eight-foot high concertina wire around it. All the bases around here look like WWII POW camps... I quickly realized that we were in a country at war."*

On his third day in Vietnam, he wrote,

I have never been so scared in my entire life—never... Mortars scare the hell out of me because you can't fight back, and you don't know where the next one is going to hit. I'm fine and intend to stay that way (MORTARS permitting). Don't worry about me—time flies when you're spending money and having a good time.

On December 27, the day of Bob Hope's visit to the troops, the show was interrupted by a firefight. Most of the audience turned to watch the gunships (helicopters) shoot rockets into the woods and wondered which unit was out there, under attack.

It only took Bill a week to declare, *"There is no glory or honor or even pride in being here. It's hot and muggy, poor resources, the people aren't appreciative, and we're not accomplishing anything."* And, once again, with a touch of ironic humor, *"I mean, I've got nothing against wars, but it just isn't any fun here."*

He sent a return address once he reached his assignment on January 1, in the 6th Battalion of the 31st Infantry Regiment, 9th Infantry Division at Fire Base Moore, operating four miles from Dong Tam in the Mekong Delta.

He became a combat platoon leader in a rifle company, leading approximately thirty-five men. He and his platoon sergeant led their men on offensive and defensive operations from the firebase, often as part of a more extensive operation. The sounds of mortar fire interrupted his voice on the tape he recorded on the days he did not write to me. I would eventually receive them and listen, never quite letting myself feel his fears, for he often included words that dismissed his anxiety. He sent a detailed account of a long day and night on patrol in mud, rice paddies, and canals, how they commandeered a small hut for the night while listening to rockets and machine gun fire and shivering in the cold from wearing wet clothes.

He included the number of GIs killed (eight) and wounded (seventeen) in the three platoons out that night, as well as a perhaps exaggerated body count of Viet Cong (eighty-four).

As long as the letters kept arriving and no assistance officers came to the door to bring bad news, I maintained my innocence and believed he was safe.

Meanwhile, still waiting for an address, I filled two or three pages and placed them in numbered envelopes daily. I made the most of a melancholy Christmas Day, telling him how much I missed him and painting a picture of our first Christmas next year with our son or daughter.

I detailed reports from my visits with the obstetrician and tales of the daily interactions with his mother and little sister. We were still getting acquainted, and I was being treated as a guest, with unspoken expectations and an agreement that I would use my meager high school bookkeeping skills to help sort out Jayne's business finances.

Alone, on an afternoon with Holly in school and Jayne at an appointment with a client, I began to explore her correspondence files. Not only were carbon copies of professional letters filed, but copies of typewritten letters to Bill were kept in another folder. Curious, I opened that file and read her letter to Bill in July after he had called to tell her of our revised wedding plans and that I was pregnant. She wrote to her son, asking him if he was sure he was the father.

A carefully revised version of that exchange was handwritten to my parents after our wedding. She had met me and my family, approved of me, and could admit her misgivings, beginning with her concerns over his meeting "a girl near an Army camp."

She wrote:

The only regrets I have are that when he phoned in March and April and said he wanted to get married, I urged him to wait, forgetting how much you need the one you love and how unimportant the other things (that parents worry about) are when you are deeply in love.

Before I left for Washington, Bill phoned and told me Bonnie was pregnant. All I could say was, "Bill, you must have loved her very much." Bill said, "Oh, Mother, I do, I do!" That was enough for me. And when I met Bonnie and her family and friends, all my doubts faded away.

If Mother had not shared Jayne's letter with me, I would have been more devastated by the letter I found in her files. I was confused by the love and acceptance that seemed to be exhibited around me. Her earlier sentiments from the summer were forgotten. I was hurt and angry and defensive, determined to be my best and purest self. I now felt I had to prove myself worthy of her respect, somehow to erase the judgment and suspicion I read into the page. Like my mistake of eavesdropping on the college switchboard, it was an action I could not undo, a sense of shame I had to hide, a story I could not tell yet remained under the surface of my psyche. Throughout the remainder of my life, my reaction to anyone judging me on first impression continues to be fierce, filled with pain, defensiveness, and anger.

My melancholy of Christmas Day away from my family increased, and longing for the familiar and loneliness made their way into letters and my teary voice on New Year's Eve. I was sick with a cough. Before I fell apart, I offered many declarations of my love and hopes for our future and our soon-to-be-born baby. I shut off the tape recorder and sat to write a letter to my parents:

New Year's Eve, 1968

Dearest Mom and Dad,

The last few hours of this eventful year are passing quickly and quietly without Bill here to help celebrate; my mind is working overtime. I've come up with a group of thoughts that I would probably be too embarrassed to share if I were with you—but since I don't have any place to demonstrate my feelings these days, this is my only means of expression.

Ever since I got your Christmas letter, I've thought about the events of 1968 and how they changed me. I think it's essential to let you in on that change.

I started the year out thoroughly confused about my purpose in life and what course I should take to find it, but ultimately determined to find out. School courses changed, my relationships with friends were altered—including a re-evaluation of John and his place (nowhere)—and I was still confused. Then, in March comes Bill, and a new dimension of trust and confidence opened up. As you could tell from my early letters about him, I could relax enough to be myself and to see what that self really was. Also, I was discovering real clues in relationships, which replaced phoniness and pretense—but not without a struggle. I was really happy those days, and I knew you shared my happiness.

From the beginning, there was a value and meaning in our love that I'd never known before, and even through our disagreements, the relationship was something I couldn't give up.

When I told you about our getting married, I had to emphasize our happiness because I knew that's what you wanted most for me. Maybe it was a bit of poor judgment not to tell you about the baby, but I couldn't place all the importance on it because then I would worry about our chances. We were torn between our new responsibility and the problems in our relationship. We just had to concentrate on making the relationship happy, so the wedding wouldn't seem like a painful remedy.

Well, it worked out. We stayed happy with each other until August and then could begin to think about our future responsibilities.

The relationship has grown steadily more secure since then. As Bill said when he was last home, "I knew we would be happy when we got married, but I never would have guessed we could get along so well and have so much fun."

I am trying to say this: I am happy and believe I've found the most stable happiness I've ever known. I realize that marriage and life have innumerable ups and downs, but I am prepared with enough love and faith in my husband and confidence in myself to conquer the hard times and come out on top.

1969—or at least 11 months of it—isn't going to be the easiest year to stay happy in—at least not a "smile-on-the-face, twinkle-in-the-eye" happy. I'll have to rely more on the deep down inside happiness I feel, and I'm sure I can do it— with a darling new baby to love and care for and discover the miracle of life with—and with my heart centered on Bill's hopes and dreams for our future.

My college friend, Ruthi, says that marriage and family are some people's way to happiness and that all the education, money, or travel in the world can't replace it. So, again, I am grateful for Ruthi's happiness formula, and I couldn't agree with her more.

Although I haven't gotten around to saying it, this letter mainly expresses my gratitude to you for letting me discover this new happiness. If I hadn't grown up happy and believing it was the way I should feel, I would never have sought so hard to find it. So, thank you for a happy childhood and for instilling my faith in people in me. Without it, I never would have confided in Bill, nor would I know how to love him so well.

Right now, happiness must be looked for and requires more effort than usual, but this year may be my big chance to grow up and discover how independent I really am.

1969 will be your first year without your "baby" home, while at the same time, your baby is discovering what babies are like. Your involvement with children's problems and their painful growing up is ending, and mine is just beginning. I pray I can manage as well.

Love Always,

Bonnie

Halfway around the world, Bill was also in a contemplative mood. On New Year's Day, he wrote a letter without any details of the daily nightmare of living in a war zone:

1 Jan 69

Dearest Kitten,

It is now 9:45 pm. I made a tape earlier this evening, but it didn't say much.

I went down to the club and listened to some Simon and Garfunkel songs. I got nostalgic.

I think my happiest thoughts of the best times were when we lived at Fort Lewis. That's the way things should be. We were on our own, and I think we both did a good job at our jobs. It was always so good to come home to you and a good meal. Just small things like making sandwiches and picking me up at work. Things were just the way they should have been. If we had been making the money we are now, what a time we would have had! Certainly, there were a lot of things that went wrong—no carpets, a lack of money, Max running off—but I think just the two of us being together made things right. I was able to do little things like put an antenna up and you did the things that made you feel like a wife; shopping, ironing, and fixing meals. I know it's a lot different now. You missed being able to fix my meals when (we were in Springfield) and I missed being able to be the sole provider. Next year is ours! We'll get our house, with carpets and it'll be a home with the three of us (maybe four if we get Max). You being a full-time wife and mother and me being able to do things around the house and when I'm not home I'll be happy knowing that at the end of the day, you'll be there just like before, with a meal ready and telling me about what happened to the baby or what the dog did. I always looked forward to coming home when we were together. God, I love you kitten. I couldn't love you or miss you more than I do. You're constantly on my mind. Put up with the situation for now. Take advantage of it. Remember the past and think about the future and smile.

I love you, Kitten,

Bill.

Bill's best friend and another classmate came to see me at 11:30, and we drank a toast to 1969 —all praying that it would bring better things.

Other than that, the New Year snuck up on us quietly without celebration. I looked cautiously and optimistically forward. When Bill's letter came on January 5, he included a reliable address. The packet of letters and a full one-hour tape of my nightly musings was sent off. I had no idea the journey they would need to travel to reach him.

CHAPTER 13
BEAUTIFUL MUD
JANUARY-MARCH 1969

I CAN'T TELL you now how I received the news. There was no telegram or knock on the door with an enlisted soldier and an officer asking to speak to me. I am guessing it was a call from the Red Cross and a kind voice who told me my husband had been hit by shrapnel from a grenade, when and where it happened, and that his location now was in a hospital in Dong Tam, Vietnam.

I do remember the date. It was January 15. Two years from the date Bill enlisted in the Army. One year from the day that he had been commissioned a second lieutenant. The day he was promoted to first lieutenant. It had only been a month since he had left Pennsylvania. All markers of time, while I was waiting to become a mother, uncertain about what lay ahead.

That day, he led his platoon on patrol, the third man back in line, entering a thick wood line off a canal. The point man, first in line, hit a booby-trapped "daisy chain" of grenades, with the first wire setting off one grenade after another. There were three explosions. Bill was hit by shrapnel from the third grenade. The first two men were severely wounded in their legs, and Bill was hit in his upper left arm.

He had been able to walk over to the "dust-off" chopper while the other two were carried on litters. They were flown to a hospital, a maze of connected Quonset huts with cement floors, big fans, plenty of light, and good food. When he told the story, he joked, "The place wasn't half bad.... I got wounded in the arm I seldom use."

I also remember the thrill of hearing his voice when he called three days later. He was elated that my letters had finally reached him, following him from his base camp to the hospital—eleven from me and several from other family and friends. It was uncertain where he would be sent to recover or how long his recovery would take—maybe Japan, maybe two or three months. We hoped he wouldn't be returned to the rice paddies immediately. We were thankful for his million-dollar wound and our spirits lifted.

When the call ended, Bill had more to say, immediately writing another letter, looking forward to our life together after Vietnam, furnishing our home, and creating a family:

> *About being a parent—I'm looking forward to it. I've done an awful lot of things in my short life—met a lot of people—been many places, and know the pitfalls one can fall into. I'll love our child so much, he'll grow up spoiled, with an open mind and goals over the rainbow and a craving to know everything about everything.*

> *Our daughters will be able to shoot straight, and our sons will be able to cook and sew for their own good, and they'll have a love for animals and their parents who are wise and omnipotent.*

How naive we were, optimistic and in love, believing the universe would always be safe and we would always have each other. How much we had yet to learn, truly experience, and confront disappointment. How lovely it is now to remember our innocence.

Less than a week after Bill's injury, he had been in four hospitals. After he was airlifted to Japan, he shared another reflection, saying that he had forgotten about the concussion of the exploding grenades and all that followed, except for the hole in his arm and how he could not feel or use the fingers on his left hand. As in the first letter he wrote on the airplane that took him to Vietnam on December 19, he shared his most profound thoughts when he arrived in Japan.

> *21 Jan 1969*

> *Dearest Kitten,*

> *Today, I saw the outside world. It's rainy and cold, and I miss you terribly.*

> *We boarded an Air Force jet early this morning and took off from Saigon at 0930. At 3 pm Japan time, we landed... and there it was—52 degrees*

of rainy, cold weather, filled with people who were busy going places or going home. There was grass and cars, all on the wrong side of the road. There were gas stations with people making a living and going home at night to their families. There were people wearing clothes— coats, hats, ties, and carrying umbrellas, and Wonder of Wonders, they all wore shoes. There were traffic lights and even traffic jams. I even saw Mustangs with fog lights; imagine that, fog lights.

Houses made with foresight, taste, and appealing to look at, with little gardens and walkways: no rice paddies or canals or mud. There is a distinct difference between mud and mud. There's mud made by rain in a garden, the mud in a bare spot on a lawn. There is a friendly mud that one can put up with. But there is a different kind of mud also. There is a mud that covers a country, a mud that is a way of life for some, and for the invaders, a gaping hindrance filled with diseases and the stench of human feces. A mud you live in because you're liberating a country for people who don't care or don't know.

But today, that mud was gone, and it was the mud that barely dirties your new shoes or that young children play in. Even mud can be beautiful to weary eyes if it's the right type of mud.

I saw cement being used to build homes and buildings where people will someday live and work; some will bring up their families and make their fortunes in these cement structures. Not cement being mixed for protection from mortars or bullets or as fortresses for machines or guardposts—no, it was peacetime cement.

On we drove, and what?—no barbed wire with its rusted points? No sandbags, no weapons, no steel pots or flak jackets? Just people going places, buying, selling, and living as people should. There should be no wars—especially the kind we are currently engaged in. If I honestly felt I was helping someone or had won something for someone, I'd have a different attitude, but as it is, we're just fighting and killing and being killed, and at the end of the day, what do you have? Have you won a hill or bunker complex or beaten the enemy out of the paddies, into the wood line, or out of the country and into Cambodia? Well, what about after a week, a month, a year? What do you have to show? Did you win the hearts of the people by giving them medicine and medical attention? If you did, why didn't they tell you where that booby trap was in their rice paddy that killed two of your friends?

The people who are the real people of Vietnam are those that work the land and have no government. They need none. They are born to a family of farmers; they work the land, marry, and take over the land, not improving it or making deals or thinking about Presidents. No, they make enough to feed themselves and to sell a portion to someone else for something in exchange.

The government can't get to those people. They can't read or write. They don't read papers, there are no roads or canals to get to them, and they would have it that way, for what can the government tell them? These people watch the GIs with one eye and the Viet Cong with the other, not caring what happens. Then, why are we here, and what are we accomplishing? Ask someone else. I'm just doing a job like everyone else here. After 365 days, my job will be done, and like everyone else who returns home, I'll forget about it, trying to erase a year from my mind because I helped no one, killed a few, and watched many die.

In a month or two, I'll be back in the mud of a country, doing my job, with no more feeling than to do the job assigned to me, do it well, bring back those I take out, and bring back myself.

Bill

The letter had not arrived before I learned the doctors in the military hospital in Japan felt Bill would receive more appropriate treatment for the numbness in his arm and hand if he were sent back to the States. With three weeks to go and as much as I was done with pregnancy, I began to pray I would not give birth until Bill came home. With the news of Bill's return, I realized I had subconsciously been trying to postpone living for a year while focusing energy on loving and caring for a new baby.

I had worked hard to have a mother-daughter relationship with Jayne, but the only thing we had in common was our love for Bill and our shared concern for his return. I missed my mother. I was filled with anticipation, yet tried to hide nervousness, unsure what would come. Without words, Jayne understood my need for my mother to be with me and again offered hospitality, inviting Mom to come once the baby was born. I let the idea of happiness begin to grow.

Jayne capitalized on her work on the election campaigns of Pennsylvania Congressman Williams. She made some phone calls and on Wednesday, January 29, Bill arrived at Valley Forge Medical Hospital, only an hour from Springfield. Jayne and I went to visit him. He still had thirteen steel sutures in his arm, but we were surprised and delighted by how much he could use his arm and hand. He managed a weekend pass and returned to the hospital on Monday only to have the sutures removed. His assignment was officially Valley Forge, but he was given thirty days of convalescent leave.

Bill visited friends the following Saturday night, and I was surprised with a baby shower given by six young mothers, the daughters and daughters-in-law of Jayne's friends. With ten children between them, ages three and one-half years to two months old, and one (besides mine) on the way, I surmised they might be Roman Catholic since, in the 1970s, birth control was considered a sin. Jayne later confirmed my assumption. The conversation turned to their experiences of complex labor and delivery and incompatible blood types—enough to increase my apprehension. Still, it was a kind and loving gesture, planned before anyone knew Bill would make it home for our baby's birth.

We were ecstatic to be reunited, yet Bill got to experience the worst part of my pregnancy, those last two weeks of my bulkiness. February 9 was my official due date. I began to have stomach cramps and thought I must be in labor. We went directly to the hospital, where I spent a few hours and was sent home, embarrassed to know so little about childbirth and what labor felt like.

Another week passed. Sunday morning, February 16, I awoke at 6 a.m. to true contractions. By 7:30, they were five and six minutes apart. This time, we called Dr. Trout to get the go-ahead to head to the hospital. Upon examination, he thought I might be there until early the following day. By noon, the contractions were four minutes apart, but Bill had gone home because the doctor said it would still be a while.

At three o'clock, just after he delivered someone else's baby, the doctor left, saying I wouldn't deliver until ten that night. I was given a shot and a couple of pills and fell asleep for two hours. At 5 p.m., I woke up quite suddenly (Did my water break? Were the contractions more painful?).

The nurse called Dr. Trout, and at 6:10 p.m., Samantha Jayne entered the world at eight pounds, eleven ounces, twenty and one-half inches long. Even at the time, the experience was surrounded by fog as I was sedated and given a saddle block. I felt no pain in my lower extremities. It only came when the drugs wore off.

Samantha and I stayed in the hospital for three days. Bill came to visit every day. I settled back into my mother-in-law's routine and advice for caring for a newborn, missing my mother but knowing she would meet her granddaughter soon enough.

Bill was given another month's leave and requested orders to return to Fort Lewis. He hoped to get his previous job back, but most importantly, he referred to it as "going home" since it was where our life together began, and it was where we established our first household.

I flew to Spokane with Samantha and went home to Edwall and the farm. Bill oversaw loading our household goods, said goodbye to his family, then drove back across the country on a shorter and quicker route than the one we had taken on our honeymoon five months earlier. He covered the 2,750 miles from driveway to driveway in three and a half long days

CHAPTER 14
THE LIFE I WANTED
APRIL-DECEMBER 1969

W E WERE APPRECIATIVE of some military privileges. Whenever orders were changed to a new duty station, the government contract with a moving company brought them to your door (often Bekins, as I remember). They tagged and inventoried, wrapped and packed everything, loaded the truck, and delivered it to your destination. Many of our few belongings had been left packed in boxes, stored in the garage in Springfield, although our stereo console was prominent in our bedroom.

We added to our inventory—baby furniture, a newly purchased black vinyl couch and chair from JC Penney, Bill's childhood rocking chair, and an old toy chest covered with a new coat of paint.

While in Portugal with the Merchant Marine, Bill had purchased a pendulum clock with plans for his future home. He carefully removed it from his mother's wall and packed it in its original crate, which had helped it survive time on his ship. Jayne insisted we take a wing-backed chair, badly in need of a slipcover, believing I had the skills to sew one up.

I left Springfield behind, taking the plane journey with a one-month-old infant in stride. While Bill drove across the country, reported for duty at Fort Lewis, and secured a place for us to live, I enjoyed my time at the farm, quietly getting accustomed to motherhood and documenting the wonder of a rapidly growing infant. I had as much to learn about babies as I did about pregnancy. I marveled at the red fuzz on her head as it evolved into fly-away auburn hair and how the blue eyes at her birth became hazel. I delighted in her smile, like her father's, and sent my mother-in-law the pictures we took in case she still had doubts about her parentage. My opinion was Samantha was the most remarkable, precious, lovable, adorable creation ever.

Barely a year after I introduced Bill to my grandparents, I went to revisit them. At 93, my grandfather was ready for us when we arrived.

As soon as I entered the door, Grandpa Green said, "Bring my youngest great-grandchild, Samantha Jayne Chandler, over here and let me hold her."

I did as he asked. Mom and Grandma joined us on the couch for a four-generation snapshot. My grandmother smiled as she told me he had her write out the name and studied it so he wouldn't forget. She added, "Samantha is our seventieth descendant."

Bill secured an apartment in Lakewood, reported for duty, and called me frequently. He described the apartment he found—three bedrooms, a carpeted living/dining room, large cedar closets, a kitchen with a breakfast nook, a washer and dryer, a carport, and a fireplace. It was ten minutes from Fort Lewis and directly across the street from Villa Plaza Shopping Center. The yard was fenced, and Max would have a secure place if Bill could not take him to the field.

My parents were relieved to see Max go. He had been a handful, emptying the kitchen garbage can if he was inside for the night and, when left outside, carrying whatever object he found in the barn or machine shed to the front porch. Our rental agreement stipulated we were to have no pets, but we were willing to chance the landlord would not show up.

Setting up our second home met my childhood expectations. I sent my husband off to work, cared for our daughter, worked in the yard and flower bed, pushed Samantha in her stroller to go window shopping across the street, did our laundry, tidied the house, made dinner, and welcomed my husband home at the end of the day.

I had time to read, sew, and write letters home. We were near enough for my parents to come for visits. We had friends come over for a barbecue (on the neighbor's grill), played lots of badminton, and went to the nearby Point Defiance Zoo. With some frustration, I met the challenge of learning to make a slipcover for a wing-back chair with curved arms. I was diligently proving myself as a homemaker for my husband and his mother. Every weekend, it was a ritual to wash the car.

Living in Lakewood limited our social life and influenced our decision to request housing on post. Bill spent five to six days a week at work while I spent those days at home. He wanted to be home on the weekend, and I wanted to go somewhere. We infrequently socialized with two military couples and expected to meet more like-minded folks if we moved. After four months, I was packing my red luggage, and friends with pick-ups came to help.

In July 1969, we returned to Fort Lewis, moving into quarters nearly identical to the ones we left the fall before. We had a three-bedroom unit. We gave up the carpeted floors and needed to buy a washer and dryer, but moving was a practical financial decision. The $120 housing allowance had not stretched to pay rent, all utilities, the phone bill, and the weekly expense of a full gas tank. Our grocery bill had been higher because it was more convenient to pop over to the local market than to buy goods at the commissary.

Five address changes in one year were nothing I enjoyed, but I told myself I would get used to it. Jayne's newly slip-covered wing-back chair and Bill's Portuguese wall clock found prominent places in our living room in each move. I also removed the white eyelet canopy from my childhood bed with Bill's urging.

Soon, Bill was inviting high school classmates over for dinner—Springfield boys who had been drafted into the Army, joined the Infantry, and ended up in Washington State for basic training. Monopoly game nights with Tim and Felix ensued. We were confident enough and comfortable with our newly furnished home to host a cocktail party for four couples and the Battalion Commander—a lieutenant colonel who bored us with war stories and tales about his classmates at West Point.

I had access to our car most afternoons, allowing me to show off our daughter to Janet and our sorority sisters at the University of Puget Sound, only fifteen miles away. After a teary start, I began to trust the caregivers at the nursery daycare on post, attended Officer Wives Club functions, and ventured alone to shop at the commissary.

Max did not fare so well. When out in the field with Bill, Max nipped at the legs of several soldiers. It meant Max needed to stay with me, either in the house or tied up outside. Let loose, he chased cats or ran away, getting us in trouble with our neighbors or the military police. He wouldn't return when we called him because he knew he would be tied up again. Tired from work, Bill no longer had the energy or interest in playing with him. In Max's confinement, the final transgression was chewing up a second pair of my shoes and pooping in our spare room. We found a better home for him, where he would have thirty acres of land, horses, other live-stock, and children to keep him company.

My life happily echoed my mother's life on the farm. When company was invited home for dinner, I was expected to offer hospitality and a tidy house. I let Bill lead the way as decisions were made about which car we owned and where we lived. This man I married resembled my father—responsible and in charge of our life together. While I was grow-ing up, the years the wheat yielded sixty bushels to the acre, Dad had cash in hand for a new car in the fall. We would sit around the dinner table with General Motors' promotional material, discussing the new exterior colors for the year, and then order the car for delivery in a few months.

Only once, while on a shopping trip to Spokane thirty-six miles from home, did Dad buy a salmon-colored Bonneville Pontiac off the lot. He drove that new car to the front of The Crescent to pick Mother up from shopping. She took the surprise in stride but frequently explained to oth-ers that she did not care for the car's color and had no say in picking it out.

Car dealers within fifty miles of military bases were happy to offer credit to soldiers. After we married, Bill's 1965 two-door GTO was traded in for a reliable four-door 1968 Dodge Dart. It took us across the country and provided independence to my life, allowing me to drive myself to doc-tor appointments while in Springfield. We decided the Dart was too small for a dog and a baby since I had to hold Samantha on my lap and give the back seat over to Max. We replaced it with a 1969 light blue Plymouth Road Runner with a darker blue vinyl top and horn that actually went "beep, beep."

Bill couldn't understand why I didn't immediately write to my father about our purchase. Maybe it was because I couldn't drive it until Bill patiently taught me how to drive a stick shift. In 1970, we decided it was necessary to own two cars. After buying a new green VW Bug in March, we purchased a used silver 1966 Thunderbird Convertible in December. In 1971, after the VW was sideswiped, we invested in a gold Volvo Sedan, including payment insurance, intending to keep it until it was paid for one way or another. Having a perpetual car payment seemed normal.

Besides Bill and Dad having lengthy conversations about cars, they had an unspoken, more profound connection. Each grew up after losing their fathers. My paternal grandfather died from gangrene and sepsis after catching his legs in farm machinery. Dad was fourteen at the time. When

Bill was twelve, his father died suddenly from a heart attack. Raised by single mothers, they learned responsibility early, grew to honor and respect women, and understood sacrifice. Financial circumstances limited their options for higher education. Most importantly, the universal expectation, "Boys don't cry" meant they both became masters of stoicism. I reflected that stoicism at every turn and did my best to honor Bill's admonition to "Be brave."

Rank also has increased responsibility. During his former assignment at Fort Lewis, Bill was an instructor, teaching classes and training troops for Vietnam. He felt useful because he knew he was doing a good job. As a first lieutenant with combat experience, he became the Operations Officer in charge of all the instructors and was responsible for their lesson plans. He didn't like the paperwork, the reports, or the need to correct others' mistakes. He was busier and happier when, within six months, he became a Company Commander with two hundred trainees under him, completing their final nine weeks of Advanced Infantry Training before receiving orders for Vietnam or, in some cases, returning to their Army Reserve units at home.

He worked long hours, spent the night in the field once a week with the troops, and frequently counseled those with personal problems. Because our future post-military life was often on our minds, Bill enrolled in night school, taking college courses through the University of Puget Sound. His brief quarter at Millersville State College in 1967 had given him only eight credits, so he had a long way to go toward a degree but planned to take equivalency tests in general knowledge subjects. I began to talk about returning to school full-time if he returned to Vietnam. I would then have only one year to finish so I could someday teach to bring in additional income.

We lived with the assumption that my income would supplement the chief breadwinner. Bill genuinely desired a life of taking pictures and raising dogs and children, but he always knew he had to find a means to provide reliable financial security.

While Bill was still in Pennsylvania, recovering from his wound, an insurance agent and colleague of his father's had come to offer Bill a job. We believed in a time of opportunity and prosperity ahead of us, but I knew (and hoped) my place was in the home until our children started school.

Eventually, the doctors at Madigan Hospital reevaluated Bill's injury. Orders dated 19 June 1969 read:

Medically qualified for duty with limitations

Ulnar Nerve Injury

No assignment requiring prolonged handling of heavy materials, including weapons.

No overhead work, no pull-ups, no push-ups.

His mobility and control of his left hand had not improved, nor had he received any physical therapy. Independently, he exercised it constantly. Surgery was considered, but there was a risk of it being more harmful than beneficial, permanently disabling his hand.

Bill read the orders as making him no longer qualified for the type of duty the Infantry is involved in, especially handling weapons. These changes were a blow to his self-image. In elementary school, he adopted the nickname "Bret Maverick" and called his best friend, Tim, "Bart," modeling their persona on the TV show. Being a gunslinger in the Wild West appealed to his imagination. He had been handling guns since he was a teenager and lettered on his high school Rifle Team.

Our love-hate relationship with the Army continued to grow. Whether to stay in or leave the service weighed on our minds. The plus column included medical benefits, a reliable, steadily increasing income, job security, and educational opportunities. A second tour in Vietnam loomed on the opposing side.

He requested a branch transfer to Military Intelligence with its opportunities for more interesting and eminently safer assignments. On his request, he also stated he would "Volunteer Indefinite," meaning he would stay in the service for an unspecified amount of time. He could get out after one more year, thus meeting his four-year obligation by January 1971. I understood it to mean he would stay in the military as long as he wanted but could also leave when he wanted.

CHAPTER 15

WHY NOT OKINAWA?
1970

WHILE WE WAITED for word on Bill's transfer to Military Intelligence, Christmas came. Samantha had taken her first steps at nine months of age. At the same time, she finally decided that her daddy was all right, and her face lit up when he came home. He had a light week without troops and spent more time with her, deciding she was all right, too. Before then, she was strictly a momma's girl, which kept me pretty busy when she needed attention at dinner time. Before, she wouldn't stay in any room unless I was there, and now she was content to watch TV with Daddy.

When Bill left for Vietnam the first time in 1968, it was one week before Christmas, and I was living with his mother and sister. We shared a dream of spending our first Christmas with our unborn child, putting up a tree, and celebrating in our own home.

In 1969, we faced the challenge of setting up a tree and keeping it from the curiosity of inquiring hands and ten-month-old Samantha's insecure footing. Our money-saving solution was to cut a three-foot tree off military property in the dark of night, buy our first strings of blue lights and two boxes of green and blue glass ornaments (all that were available at the PX), and set the tree on a table.

Samantha stood underneath it and watched it but never touched it. One night, we were awakened by the sound of it crashing to the floor on its own, smashing half the ornaments. In the morning, we re-placed the tree on the table, discarded shattered ornaments, and secured the tree, a little worse for wear. What mattered most that Christmas was our child's delight in the wrappings her new toys came in along with our togetherness.

In January, I attended the ceremony where Bill was promoted to captain. He came directly from the field in his fatigues and claimed it was "no big thing," but I thought it was significant. It was January 15, 1970, ironically the anniversary of his shrapnel wounds in the rice paddies of Vietnam, which was the furthest thing from our minds.

A new door then opened. Bill received an unexpected set of orders: Permanent Change of Station assigned to "Fort Buckner, HQ USURY'S (Okinawa) Reporting 5 March 1970." We shared our news with colleagues at holiday parties, and the reactions were positive and encouraging.

"Oh. You'll love it!"

"Wish that's where we were going!"

"There's so much to see over there."

"Oh, you'll love to go shopping."

We began planning another move with thirty days leave before Bill would report for duty. We would all stay at the farm with my parents for February, and Bill would go to Okinawa and wait for housing on post. It might take six weeks or as many as twenty weeks before Samantha and I could join him. The scheduled length of the tour was two and one-half years, but we knew several things could alter plans—a branch transfer to Military Intelligence being one of them.

We had acquired more stuff, and our belongings needed to be sorted into two lots, one for storage and one to be shipped. Bill was completing the final weeks with his last training group, so I was more involved in the moving process than before. Our furniture, some books, and all the paraphernalia we could live without would be stored. Our washer and dryer, dishes, bedding, towels, cooking utensils, typewriter, sewing machine, Bill's tool chest, etc., were to be shipped. Our quarters would be furnished.

Maybe it was the number of moves or the stress of this one, but again, the subject of getting out of the service and having a permanent home was revisited. Settling into a permanent home seemed an excellent idea. "Maybe just three or four years more," Bill said. The dance continued.

Samantha got sick, vomiting in her crib the night before the movers came. I worried about her all day while the movers were there. She still wasn't feeling well that night, so we decided she and I would fly to Spokane and leave Bill to finish cleaning and stay in the Visiting Officers' Quarters. She was sick with diarrhea for three more days but did not act like she felt terrible. By Monday, her appetite returned.

Then, to keep the stress going, while Bill and I slept in she fell against the cold air return Tuesday morning while carrying a glass bottle, cutting her hand from her thumb to between her middle and ring fingers. Mom felt responsible but eventually asked me why I only had glass bottles. Samantha's first medical emergency trip was to my childhood doctor. He closed the wound with four stitches.

Bill had arrived the previous Saturday night, shaken up. His training officer had taken over as CO (Commanding Officer) two days before. He told me three troops had been in the arms room—the guard and two others. One was sitting by the window when the third grabbed the guard's M-16, loaded it with a magazine of live rounds, and set the trigger on automatic. Messing around with the weapon, the trigger was touched, shooting three rounds. Two of the three rounds hit the guy on the windowsill. He died on the way to the hospital. He was only nineteen years old. It was a tragic, terrible waste—especially since there was no palatable cause or explanation—much like casualties of the war we were fighting in Southeast Asia. It was a sad note to leave Fort Lewis on.

We planned for Samantha and me to stay with my folks until we could join Bill in Okinawa. The Army would feed and house him, and we would save on expenses. We took three months of advanced pay to cover the loans on the stereo, the washer and dryer, and our living room furniture. We didn't want to ship a car, so we sold his beloved Road Runner and were out from under a car payment.

Then, Bill's top-secret security clearance was approved, and his transfer to Military Intelligence came through. The Army notified him of another change of orders. Instead of Okinawa on March 5, he was now to report to Fort Holabird, Baltimore, Maryland, on March 3. He had to complete a six-month Aerial Surveillance Course before we would know his next assignment. Another move, another adventure. My red luggage was ready.

Over the years, I have fought the cynicism and negativity of my disenchantment with military life. In the 1960s during the draft, the job of military recruiters was simple—convince those with a low draft number that they would fare better as volunteers. If your draft number was called, you appeared before a local draft board of community members who then began to evaluate your draft status. Statistics show that one-third of American troops in Vietnam were collected by the Selective Service System. Most drafted soldiers were from poor and working-class families.

I imagine, but cannot know, how much Bill was influenced by a sense of duty and patriotism, shaped by the role model of a father who'd fought in WWII, formed by the circumstances of his father's early death, and motivated by an ingrained responsibility for his actions, such as my pregnancy. Revisiting the story of waiting for the Army to decide our next move, knowing he was a pawn in a war machine at an unfortunate time in history, I remember him as doing his best with his life and for his family. At the time, we took the waiting in stride. We cooperated with the system.

Being sent to Baltimore meant we would be only ninety miles from Springfield. Jayne offered us a place to stay until we could get settled. Before heading east, we celebrated Samantha's first birthday with my classmates and cousins, who all had small children.

Jayne had not seen her granddaughter for a year and seemed eager to get acquainted, offering to babysit for us if we wanted to go out alone. We flew to Philadelphia with plans for Samantha and me to stay with Jayne and his sister Holly while Bill went to Baltimore to report for duty and find housing, joining us on weekends.

We needed a car, and the Monday after we arrived, we bought a brand-new Volkswagen, a smaller car with smaller payments. We were basically out of debt and had a built-in babysitter and places to go. We went to movies, and when Samantha was napping, we were free to buy groceries at the commissary at Valley Forge and go furniture shopping, building our vision of a future place of our own. Bill also took me to see his high school—huge compared to mine.

We expected to stay at least three weeks. The facade of "welcome home anytime" began to show its cracks. Samantha had just started shaking her head when we said, "No, no" to her. There were lots of "no-nos" in Jayne's house and lots of frustration on my part. Jayne was eager to share her opinion about raising children, criticizing her daughter Cherie's parenting, although Cherie lived in Louisiana and Jayne seldom saw her. I held my ground and tongue, telling myself that my methods worked nearly as well. Samantha displayed her good side after the first night when she cried and cried. Jayne pronounced her the sweetest and most precious baby.

Jayne also constantly talked about being short on money. I believed it was a family matter and not my business, and it was distressing to hear her talk about it. I remained the quiet, polite, model wife, mother, and daughter-in-law I had tried out the year before.

CHAPTER 16
ON OUR OWN

I T IS MARCH 13, 1970.
  It is our last weekend before Bill reports to Fort Holabird, where he will serve as Staff Duty Officer once every three days for a twenty-four-hour shift. We want to make the most of the weekend and plan another day of exploring. We take Samantha downstairs for breakfast.

Jayne sits with us and begins to tell us that she would like us to contribute more to the expenses of staying with them. We have done something to upset Holly, and we need to be more considerate. When Jayne calls us selfish and irresponsible, Bill raises his voice, defending our actions, as confused as I am about what she means. As tension between them escalates, the shouting increases, and I begin to cry. Samantha still needs to be fed.

After a week as Jayne's guests and being criticized for our immaturity and being a financial burden, it is clear Bill, Samantha, and I needed to leave for Baltimore. From my past observations, Bill knows I felt we were a burden because of Jayne's constant comments about her financial situation. Because we are facing six months of half-pay due to the advance we had taken upon leaving Fort Lewis, there is no way we can contribute more to Jayne's expenses. Bill decides it would be best not to leave me in Springfield alone, meek, polite, upset, and defenseless. Without wanting to explain our decision, as soon as Jayne leaves the house, we pack our suitcases and leave. Bill leaves a note, writing down a forwarding address.

We arrive in Baltimore in the early afternoon. We check into a room in the Guest House, a group of bedrooms with a shared lounge, a kitchen with a refrigerator, and facilities for warming baby food. It's designed for temporary lodging while people look for apartments. Several other children are staying here, including two babies about eleven months old. Samantha is busy and happy, although once again, she has a bad cold and cough in the transition of moving.

We found out the apartment Bill had put our name on wouldn't open until the middle of April. We leave Samantha at the nursery to go apartment hunting and find a two-bedroom apartment only five or six blocks from the main gate. We feel lucky that our monthly rent will be $119.50, plus gas and electricity. Our move-in date is three days away, with or without furniture.

I write a letter to my mother the next day, admitting my despair and trying to sound reasonable and make sense. We believed we were behaving like grown-ups, yet realized we relied heavily on our parents as we felt it necessary, because they offered and because we could.

MARCH 15, 1970

I have written home, but my letter is still in transit when Mother calls. There is one phone in the Guest House, and on Sunday afternoon, I am called to the phone.

"Hello?" I say.

"Hello." I hear her voice, and the tears start to fall.

"How did you get this number," I ask.

"I tried to reach you at Jayne's," Mother says. "She told me you had left, and she told me how to reach you."

I don't know how to respond, nor how Jayne conveyed our leaving, and I am overwhelmed by yearning for my mother's comforting support. Years of being tuned in to each other's moods and guessing about feelings by sight fail us over the phone. I am glad she knows we left, yet I am embarrassed and cannot explain. Having expected to live with my parents for up to three months, leaving the relaxing comfort of the familiar, I miss her terribly. I haven't yet figured out how Bill and I are suddenly faced with feeling totally on our own.

MARCH 20, 1970

After six days in the Guest House, we move into the apartment with nothing but our suitcases filled with clothes and Samantha's toys. Bill runs into classmates he had in Officer Candidate School two years earlier. One gives us some pots and pans, two TV trays, and two folding chairs to help us get by. The other couple offers a mattress and some bedding for us to sleep on the floor, along with a crib for Samantha.

There is no heat. The gas company did not come when they promised they would. After waiting all day, Bill calls the manager at about 5 p.m., and they apologize, sending someone over around 6 p.m. to discover part of the furnace is missing. The building owner, a janitor, and two maintenance men come to repair the furnace. By 7:30 p.m. the furnace is workable, but not the thermostat. We have to connect two wires when we want heat and disconnect them when we don't.

Another problem is the stove needs cleaning. We complain for a day or two until we are told the cleaner has been fired for doing such a lousy job. They offer us five dollars off our rent for the inconvenience and bring us oven cleaner and Spic and Span. We tackle a considerable accumulation of grease on top of the refrigerator, the walls, and the side of the stove. There is nothing we can do about the dirty floors.

But, we tell ourselves, the place has been freshly painted and looks clean. We survive by saying, "We'll only be here six months, and once we have our things…. Well, things must improve because they can't get much worse."

Meanwhile, I feel terrible about the impasse with Jayne. I believe I was at fault, guilty that I caused the rift. A lot can be settled if we talk things over.

I suggest to my husband, "Maybe we should have stayed longer, discussed our differences, and tried to make things work."

Bill, the strong-willed one, says, "I won't stay in a house where I am constantly reminded how much I owe that person."

That's all he says. Subject closed.

MARCH 21, 1970

Jayne brings us a lamp, two lawn chairs, and a picnic basket filled with groceries and plasticware. There needs to be a conversation about our departure, but it does not happen. I do not tell her about staying awake watching the mouse climbing along the window ledge of our bottom-floor apartment or how I have already learned to close my eyes and count to five to let the cockroaches hide when I turn on the bathroom light at night. I want to show her we are doing fine without her help.

MARCH 29, 1970

We spend Easter in Springfield with Jayne, Chris, and Holly. I feel nervous, defensive, and, at other times, very humble. Bill's relationship with his mother, based on a lifetime of shared experience, hasn't changed, and my discomfort within a new family system remains.

Jayne writes to my mother the next day:

March 30, 1970

Dear Dorothy,

Yesterday was Easter and I had invited Bonnie and Bill and Samantha for the weekend or the day, whichever was the easier for them. It turned out to be a miserable rainy day, but the kids drove from Baltimore in the pouring rain.

Bonnie and Bill were darling and came with a big yellow chrysanthemum plant and hugs, so everything is behind us now I hope. We had a lovely time, a good visit, and a nice dinner....

About 12:30 the heavy rain turned to sleet—then snow—so when they left last night we had about two inches of snow on the ground and the roads were treacherous. I sure hated to see them drive back, but they expected the movers today and Bill had to be back. They gave two rings on the phone to let me know they arrived safely.

Bonnie made a point of telling me how much she appreciated the things i took down to them last weekend and thanked me for everything. I tried to make them feel at ease and things went well, I'm sure, but it was nice to hear....

Sorry there was any misunderstanding between the children and me—but it is over now. Love them heaps and bundles—no less because I had to scold them. Take care.

Love, Jayne

In the eighteen months that follow, we will visit Springfield often. I hold my tongue as Jayne criticizes and lectures Bill, giving him excessive advice. She frequently privately tells me her opinion of his faults and expects me to agree. I realize he is immature about some things, but I am his wife and love him unconditionally. Besides, I am younger than he is and equally immature. I don't think she needs to lecture either of us about unwise decisions—time will teach us and change us soon enough. My patience wears thin. Such conversations and experiences enforce my practice of how "not to behave."

Most importantly, conflict with Jayne influenced our goals in creating the family we dreamed of—one where voices will not be raised in anger, and money matters will not become an issue that our children need to hear about or worry about.

It wasn't long before I shared this analysis of the situation in a letter sent along the Round Robin route to all my family:

We take relationships with relatives (especially parents) so much for granted that we don't work hard to show gratitude. Since they have always given us love, material goods, and attention, we expect they always will. It's okay to accept it all silently as children, but as we grow older and supposedly more mature, we should express our appreciation—but sometimes we don't.

PRESENT DAY (2023)

I was not this reflective then, but over the last fifty years, I have reflected on the Chandler family drama and my reactions. In telling this story now, I take less responsibility for what happened between us as caused by ingratitude. Parents always want the best *for* their children and expect the best *of* them.

I did not understand the stress of a working mother, a single breadwinner in her early fifties, perhaps experiencing the hormone shifts of perimenopause. Nor did I fully appreciate how parents never quite let go of their children—when children leave home, it creates a void and change of identity for everyone. We had been married for only two years, and family structures always need time to readjust. My resilience grew, and such experiences strengthened my independence. This is what I have come to know about myself:

Expressions of anger frighten me. As the youngest in my family, I learned how *not* to behave by watching and listening carefully. My siblings had left the farm by the time I was thirteen, and I had five years as the remaining child, alone with my mother. Mother confided in me her frustration with my father if he was late for supper, but she always cooled off before he got home. Nothing was said, and I was unaware of how she may have expressed her anger to him privately. I remembered her interactions and disappointments with my oldest sister's promiscuity and my brother's drinking. I learned how to react to unspoken disapproval when I saw a disappointed look on her face, no matter the cause. The family name and reputation were always an issue.

I remember hearing voices raised on the farm only once in my childhood. I was upstairs in my bedroom and listened to my father yelling at my oldest sister, Sandy. She wanted a transistor radio to take to college, which my father thought was an unnecessary expense. Sandy resorted to calling him names and declaring she hated him. Dad responded in anger, telling her she was a spoiled brat. It frightened me and stayed with me because it was so unusual.

Fear still makes me cry. For years, I wondered about the source of my fear of abandonment, yet knowing it was connected to a community picnic on the Fourth of July when I was a toddler. When I asked my sister, Susan, what she remembers, I learned that there was a time when our mother, a lightweight drinker like me, had earned a community reputation as a flirt.

On this particular Fourth of July, her drinking behavior upset our father to the point that they began to argue. They didn't want to make a scene, so they drove away in our car to air their disagreements. What stayed in my memory was coming out of the creek where I had been wading. I was crying and looking for my mother. No one would or could tell me where she was. Susan was there to comfort me. I was three. She was eight. Unknown to me, my parents struck a bargain that day—Mother would not drink so much if Dad would go to church with us. Remembering how we sat together in the same pew every week, I am aware they both kept their end of the deal.

CHAPTER 17
BALTIMORE –
AN EXCELLENT PLACE
TO VISIT

APRIL - SEPTEMBER 1970

T HE FIRST TIME I saw the musical *Hair Spray* in 2007, I was startled as Tracey Turnblad woke up and started singing, "Good Morning Baltimore." She leaves her row house on the way to school, singing about encountering "rats on the street... the streaker next door... and the bum at the bar." I smiled at the Baltimore I remembered. The fictional character (who wins a spot on a local TV dance show and then teaches 1962 Baltimore a thing or two about integration) may LOVE Baltimore, but I never did.

Living in an industrial section of town in a small basement apartment with mice and cockroaches and having no yard for our active and curious eighteen-month-old daughter to explore did not win my heart. In order to repay the Army for his advanced pay, for the six months we lived there, we got by on half Bill's salary. Because he was a student, no longer in charge of troops leading Advanced Infantry Training, we had weekends to ourselves and filled them with day trips that cost no more than a partial gas tank in our economical little Volkswagen. Before we got acquainted with the nineteen other members of Bill's class, we explored independently. We later learned that they were spending their weekends the same way.

Baltimore was close enough to Washington, DC, we drove in, spent ninety minutes viewing the White House and the Washington Monument, parked the car to walk inside the Lincoln Memorial to get a good view, watched the solemn changing of the guards at the Tomb of the Unknown Soldier, and drove past the Jefferson Memorial, which was closed for repairs.

The following weekend, we went to Annapolis to see the Naval Academy. Sailboat races there gave Bill plenty of subjects for photography. His manual focus single lens reflex Pentax camera and extra rolls of film were always with him. When my family made their way to the Washington area, we acted as tour guides, meeting my sister Susan, Dick, and their boys at the Lincoln Memorial and seeing the presidential helicopter land on the White House lawn.

We took longer excursions—west into Maryland County to get to Harper's Ferry National Park, West Virginia. The National Park Service always impressed us with its protection and recreation of historic sites. We heard the story of John Brown's uprising in Harper's Ferry, followed by a tour of Antietam Battlefield National Cemetery, with the gravesites of four thousand Union Soldiers. This began our immersion and education about the Civil War. By August, when we went to see Gettysburg Battlefield with friends who were visiting from Fort Lewis, I had started to think that once you've seen one Civil War Battlefield, you've seen them all.

Another Saturday, we started from Jayne's house in Springfield, drove to Atlantic City, and explored the Boardwalk's hamburger stands, clothing stores, souvenir shops, import shops, and jewelry auction houses with Samantha in a stroller. I felt like a little kid, having never seen anything like it, with acres of sparsely populated sandy beaches on a cool, windy day in April.

Our forty-mile drive down the Atlantic coast from there helped me understand Bill's comment about liking the Atlantic better than the Pacific. Villages and resort towns along the beach were what he had known as a kid. At the same time, my experience of the ocean began at age eight, when we vacationed in rustic tourist cottages at Long Beach, Washington, playing in the sand, hunting for driftwood, riding reluctant horses, visiting lighthouses, discovering saltwater taffy, and splashing in frigid water, while the men went salmon fishing. I was impressed but still preferred the rugged beauty of the Pacific Northwest Coast.

We concluded the day with an eighty-minute ferry ride from Cape May, New Jersey, to Lewes, Delaware, then drove two hours to the Chesapeake Bay Bridge and returned home via the Harbor tunnel to Baltimore. It was a twelve-hour day, and Samantha was angelic and well-behaved. Toll roads were another new experience for me, and though

I extolled the gas mileage of twenty-six miles per gallon, I complained about the tolls. The New Jersey turnpike to Atlantic City cost $1.25; the toll bridges to Cape May were $1.25; the ferry was $5.25; the Chesapeake Bay Bridge was $1; and the Harbor tunnel into Baltimore was $0.60 cents. We spent $9.35 for the day in tolls.

Jayne offered Bill the opportunity to take pictures for some of her advertising clients. The first client needed before and after photos of hair receiving split-ends treatment. My hair got the treatment (though I was slightly offended that she thought I had split ends), and Bill took the pictures. He even got paid for it. Without words, it may have been her way to acknowledge her son's skills and encourage him. The arrangement was for our mutual benefit. Her client paid her 125 dollars for a professional photographer. Because Bill was the photographer, he received half of that amount as payment. It did remove some of the tension in our relationship.

We had other reasons to go to Springfield on weekends. Bill's younger brother graduated from high school. One of Bill's best friends graduated from college. Bill's high school class held a five-year reunion. I was meeting more and more of Bill's friends, finding out how well he had always been liked, and understanding how he quickly attracted workmates and became friends.

Our visits with Jayne were pleasant as long as we listened to her guidance and advice. Playing croquet in the backyard and letting Samantha run around outside were nice changes from the concrete jungle of Baltimore.

Bill's classmates at Fort Holabird were near our age and, like us, on their own for the first time. Some were married, but most were single. Class picnics were held at Gunpowder Falls State Park, where the Gunpowder River flows into the Chesapeake Bay. There was swimming, badminton, softball, volleyball, and lots of food. Samantha loved playing in the sand, but I was the only mom chasing after a baby while the other two little girls, younger than Samantha, were peacefully penned up in playpens. Our apartment became a gathering place for dinner and nights of Monopoly. Bill usually won. I frequently lost. If I went out first, it sometimes meant I could go off to bed and let it be a guys' night. No one seemed to miss me, and I always woke up for good night kisses as well as other loving activities.

In August, our next-door neighbor in the apartments, Mike Book, one of Bill's classmates, had his girlfriend Susie come from Florida to stay with him until he left for Vietnam in October. Like me, she had nothing much to do during the day. She asked me if I knew how to play bridge. Though I had sat at the table, holding a place for the occasional missing member of my mother's bi-monthly bridge clubs, I only had a smattering of knowledge of how the game was played. While Samantha napped, Susie and I spent most afternoons dealing out hands, talking through how to bid and how to play. At least three nights a week, Bill and Mike joined us, and for our final three weeks in Baltimore, we stayed up playing until after midnight nearly every night. Initially, Bill was reluctant but learned quickly. He said he played for fun while I took the game seriously, wanting to learn to play well.

On evenings without Monopoly or our marathon Bridge games, Bill and I sat comfortably across the room from each other, reading. Reading was always part of my life, and I carried the image of the floor-to-ceiling, wall-to-wall bookcase my father had built on the farm. Joining the Book-of-the-Month Club provided another link to my mother's life. *The Secret of Santa Vittoria* by Robert Crichton, *The Cancer Ward* by Alexander Solzhenitsyn, *A Quiet Voyage Home* by Richard Jessup, and *The French Lieutenant's Woman* by John Fowles were added to our much smaller bookcase. Letters to my family often included comments about what we were reading and making recommendations to one another.

Bill's full salary was restored, classes were complete, and our six months in Baltimore ended with Bill's graduation, which was nothing more than receiving a certificate of completion and a handshake. We were so thankful to leave Baltimore; it seemed like an actual occasion. The move to Hillcrest Heights, a Maryland suburb of DC, was accomplished in one day. Everything was packed up and loaded onto a truck in less than four hours, and everything was unloaded and partially unpacked by the movers by six o'clock that evening. It was the fastest we ever got reorganized, and we felt that if the Army wanted to move us fifty miles at a time, it would be okay with us.

CHAPTER 18
HOME SWEET APARTMENT

A YEAR IN DC

SEPTEMBER 1970-SEPTEMBER 1971

B EFORE WE MOVED from Baltimore, we toured six or seven model homes in the DC area. We were told Bill's assignment could be up to four years. Living in one place seemed to answer our mutual desire to create a home with a yard and area for our children, plus a dog or two to run and play.

I had clipped a magazine article about interest rates and the increasing price of land and housing. It concluded with these lines: "Waiting for a drop in mortgage interest rates is not the answer. Remember, when you rent space, you own nothing; when you rent money, you can own a home." We were determined to have a home, but our budget did not stretch that far. The apartment complex we found, with a swimming pool and a preschool program for children eighteen months to six years, was more workable. I shared my excitement in a letter home:

We went apartment hunting and came up with the perfect place. We're getting into a second-floor, two-bedroom apartment with a den. It's on the outskirts of Washington and about fifteen minutes from where Bill will work. We'll still be in Maryland, on the southern side of DC. It has benefits like a dishwasher, three walk-in closets, a separate dining area, and a balcony twelve feet long and six feet wide. All the utilities are included with the rent. We'll have our own dedicated storage space in the basement. In addition, the apartment complex is enormous. There is a swimming pool, tennis courts, a community center with planned activities, plus lists of babysitters within the complex—all information published in a monthly newsletter. We'll be about fifteen minutes from Andrews Air Force Base, where I will do my commissary shopping.

Through a wholesale furniture dealer, we found affordable furniture items. We had been looking for at least a year, anxious to furnish our place without borrowing from colleagues or depending on Jayne. We bought two end tables, a coffee table, a round dining room table with four chairs, and a matching hutch with glass shelves and lights in the cabinet. It was delivered C.O.D. (Cash on Delivery). We had four hundred dollars in cash and financed the remaining two hundred dollars. Bill's classmate from our time in Baltimore returned our washer and dryer to us, but we needed to take them to Jayne's basement for storage.

We furnished the den with a twin bed from the house in Springfield. The pendulum wall clock Bill bought in Portugal found its place on the wall. Small bookcases from Bill's childhood bedroom were filled with our growing Book of the Month selections. There was just enough room for Jayne's wingback chair. It became the room for overnight guests. Our bridge partners, Mike and Susie, visited from Baltimore, and the room was set up for them the first weekend we were in our new place. That weekend was filled with bridge games, sightseeing, and hours of talking.

Jayne stayed overnight once after meeting with a client in Washington, and Bill's brother Chris drove from Springfield before he reported for duty with the Navy. Like his brother, Chris' draft number was low (number nine), which meant he would be going into the Army shortly. On Bill's advice, Chris decided that any service would be better than being a private in the Infantry in Vietnam. Bill spoke to the recruiter, telling him Chris got seasick. Chris never had to serve on a ship. His first assignment was Bermuda.

My cousin Marla, now married to John, my high school sweetheart, and Marla's younger brother Bernie, came from Fort Bragg, North Carolina, for a weekend. Bernie had a cheaper ticket if he flew back to Dartmouth College in New Hampshire from Washington, DC, rather than Durham. We spent the night reminiscing about high school basketball championships and gossiping about people we knew. Bill felt a little peeved as we dealt so much with a past he knew nothing about. I had few chances to talk to anyone from home. I was more often around Bill's friends and felt on the outside in the same way.

I wondered, *Is a little tolerance from him too much to ask?*

Memorial Day weekend, my sister Sue, her husband Dick, and their sons spent four days with us before leaving for Dick's assignment with the Air Force in Izmir, Turkey. Our apartment suddenly seemed tiny with seven of us together. Samantha was enchanted to have two cousins to play with. Bill and Dick took all three kids to the Navy Yard Museum, where they climbed over weapons and ships, and to Fort Washington, where they had room to run. Sue and I did laundry while they were gone. Bill drove with Dick to Philadelphia to ship their van, then we sent Dick, Sue, and the boys to Turkey via New York.

No matter who came to see us, Bill worried about how the place looked and what we would have to eat. I was more relaxed and ready to enjoy their visits since I spent plenty of time on housework, knew the house was clean, and figured we didn't have to impress friends or family. I was happily filling the role of a devoted housewife and competent mother.

Our apartment became a place where we welcomed new friends and entertained them. Samantha and I made friends at the swimming pool and outside on the playground. I signed up for activities at the community center. I invited the women I met for coffee, then we asked the couples for dinner. We played bridge with friends several times every week. I joined a "babysitting co-op" where members cared for each other's kids. I earned chips (poker chips) by caring for other mothers' children and reciprocated by using my chips for them to babysit Samantha. It was an excellent economic system since babysitters typically charged seventy-five cents an hour.

Merry-K, my childhood friend and bridesmaid at our wedding, was at George Washington University, living in Arlington, Virginia, fulfilling her lifetime goal as a nurse. She roomed with six other nurses. We hosted our first Thanksgiving dinner for Merry-K and her boyfriend, Ethan. I couldn't imagine Thanksgiving without family or a turkey, so I cooked a six-pound turkey, made a green bean casserole with mushroom soup and canned fried onions, baked rolls, and made a pumpkin pie. The evening ended with us playing bridge. Although everyone had to get up in the morning, they stayed until 1:30 a.m.

Given our location, it was natural for our neighbors, friends, and acquaintances to all be connected with the government—FBI, Army, or Air Force. Most of them had security clearances. All were conservative dressers, very clean-cut. Non-military government workers considered themselves part of the "establishment" and saw us as "the enemy" since the war in Vietnam was unpopular. We were also younger than many of them.

A high school classmate of Bill's, a bachelor who worked for the FBI, invited us to a party. One guest heard Bill's story about being wounded in Vietnam and called him "a regular John Wayne!" Most everyone there was unmarried and had a date. They seemed more serious, eager to make the right impression, and more likely to be in physical contact with the opposite sex. We took notice of all the physical connection, intense conversation, and dim lighting, trying to remember acting the same way three short years earlier.

Our apartment became a place for Samantha to grow. At eighteen months, she gave up taking a bottle to bed at night. She became full of copy-cat ideas, helping me unload the dishwasher, throwing scraps of paper away by opening a folding door to reach the trash can, putting clothes in the dirty clothes hamper in the bathroom, and opening every door except for the front door. She attended preschool three days a week, two hours a day. She was kept busy with games, songs, and little projects, given cookies and Kool-Aid. I quickly learned to enjoy having a second cup of coffee in a quiet home.

By the age of two, Samantha was potty trained. She came home from her preschool singing parts of various songs, often leaving us guessing just what she was singing. She watched Sesame Street every morning and loved learning the alphabet, with her favorite letters being U, X, Q, and W. When summer began, we often went to the swimming pool. After the second day of splashing around while Bill and I hung on to her, she started fearlessly jumping into the four feet of water. The baby pool, about one foot deep, was pretty tame to her. She was so tired every night that she slept very well, and I discovered that being out of the apartment for large parts of the day cut down on housework.

We took Samantha to the National Zoo. We saw white tigers, a pygmy hippopotamus, and the one and only "Smokey Bear." We enjoyed it more than Samantha, but she was more interested than the year before at the Baltimore Zoo when she chased the pigeons, or the year we took her to the Seattle Zoo and she slept the entire time in her stroller. On this trip, we took a half-hour train ride through the zoo, which she enjoyed the most.

The apartment was where we envisioned our future. Bill studied and wrote papers for his college classes, sometimes employing my typing skills. He attended night school on Monday and Thursday, taking Speech and Sociology from the University of Maryland. After finishing the fall semester with a B average and gaining six more credits, he decided to take only one course, signing up for World Literature. He took several tests through the Army College Level Equivalency Program and was given college credits in general subject areas. The Army promised to credit him for military training courses through the University of Nebraska. He had credits equivalent to two years of college, another incentive to stay in the Army, and the possibility to consider living in Lincoln, Nebraska, for a year or two.

Our living room became an occasional studio for photography jobs. Bill bought two floodlights with stands, a diffuser screen, and a large roll of white paper for a seamless background. He took a bride's portrait and headshots for some colleagues. We worked together photographing a wedding, exchanging loving glances while the Carpenter's song "We've Only Just Begun" was played. Jayne arranged another job with one of her advertisers, which netted us two hundred dollars. Bill always took advantage of facilities on the nearest Army base to develop his black-and-white pictures.

We learned he could count photography expenses as a tax deduction, which we celebrated with a solo "date-without-daughter," window shopping, exploring downtown DC, and having lunch together. The eight hundred dollar tax return came later.

Living near Washington, DC, we became guides to national monuments and battlefields. Among them was the Changing of the Guard ceremony at the Tomb of the Unknown Soldier, the Washington Monument, and the Jefferson Memorial, which became our favorite. We explored at least three Smithsonian buildings and saw the high points, including the Hope Diamond, the Dinosaur exhibit, moon rocks, and the Wright Brother's plane.

We enjoyed the beauty of fall in the Virginia countryside, driving to the Shenandoah National Forest and part of the Blue Ridge Mountains: Mount Vernon, Washington's home, and another Civil War site, Manassas Battlefield. In March, with the temperature reaching seventy degrees, we discovered a State Park in Virginia—Great Falls of the

Potomac—and went there for a picnic with another couple and their eighteen-month-old boy.

On one of our "history tours," we went to the birthplaces of George Washington, James Monroe, and Robert E. Lee—all in Virginia. Monroe had a highway marker, Washington had a rebuilt house and national park, and Lee had a beautiful mansion and plantation on the Potomac River called Stratford Hall. Lee's land fit how I believed a plantation would be: an outdoor kitchen, schoolhouse, smoke house, stables, a mill, and numerous enslaved people's quarters. The self-sufficiency of such living became apparent.

Getting away from the city and our large complex of three-story apartment buildings was refreshing. Some weekends, we went for a drive, one time ending up at the southernmost point of the Potomac River Peninsula at Lookout Point State Park in Maryland. We took the scenic route through Pennsylvania farm country after a visit with Grandma Jayne, passing Amish farmers out for a Sunday drive in their horses and buggies. Away from the hustle and bustle of DC, it was nice to see people living a simple life. It was a little like home, and the drive caused a bit of homesickness.

At Christmas, I prepared to spend the holiday in Springfield, meeting Bill's sister Cherie, her husband, and two sons from Louisiana for the first time. I knew the small house would become crowded, and relative strangers would surround me. I wrote to my family, letting them know I missed being home while trying to convince myself our new life away from the farm was the natural order.

December 1970

For each of us, our present address, the place where our husbands and children are, and the place where we work and shop and share moments with current friends is HOME. And to everyone, that big old house in the country, where we all spent so many happy hours and days, will always be "home" too. We are fortunate to have a place we can always go home and refer to as home.

We had other date nights where we enjoyed the abundance of available nightlife. A coworker of Bill's told him about a schoolteacher performing at Mr. Henry's restaurant on Capitol Hill. She sang jazz, accompanying herself on piano. When she sang "The First Time Ever I Saw Your Face," it became our song. We learned that night her name was Roberta Flack.

When Sue and Dick stayed with us, they treated us to a night with Anthony Newley and Buddy Hackett at a theater-in-the-round. Dick's loud, hearty laughter caused the people behind us to get up and leave, or perhaps it was because Buddy Hackett's jokes got increasingly off-color. Another night, Bill and I attended a live production of *Fiddler on the Roof*, the first stage performance I had seen since college.

On the third of July, we took Samantha to a *Torchlight Tattoo* at the Jefferson Memorial and discovered it was being taped for broadcast on the Fourth. The Army Band, Chorus, the Army Drill Team, and the Old Guard Fife and Drum Corps performed precisely and skillfully. The finale was a capsule history of the development of the United States flag and a salute to all the states. It was impressive, and I appreciated living in the nation's capital for the first time.

After not seeing my parents for over a year, I retrieved the largest suitcase in my set of red luggage from our basement storage area, and shortly after Samantha's second birthday, I went home for a three-week visit to the West Coast. I spent time with my folks on the farm, drove to the ocean for a reunion with both sisters and all seven of Samantha's cousins, and went to Tacoma to visit college friends. I went out for dinner one night, leaving Samantha with my parents in the motel room. As the night got later, Samantha stood looking out the window. My mother came out of the bedroom to talk her into going to bed.

Samantha replied, "Go to bed, Grandma, I'm waiting for my momma."

While shopping in Spokane, my sister and I talked to each other about buying wigs. When Samantha and I flew back to DC, I wore the wig. Bill needed a second look to recognize me. Samantha wasn't sure the man who picked her up at the airport was her father. She didn't say much until we got to the car. She wanted to play with her blocks and other toys right away.

It didn't take long, and she soon said with great glee, "That's my Daddy." She was attached to him for several days. Bill noticed changes in her looks and how much she had to say. It seemed to both of us that suddenly, she was talking all the time.

I was surprised on my twenty-third birthday when Merry-K and Ethan arrived for an evening of bridge, bringing a bottle of wine, a present, and a birthday cake. A week later, Bill and I celebrated our third wedding anniversary. In that time of celebration and reflection, we felt we were successfully navigating the waters of marriage and adulthood. We were living our lives in a way that suited us and expected nothing but more and more contentment. We had a child we adored. Bill told me he loved me more than ever.

He said, "We have been happier in the last six months than at any other time."

I agreed. It was a great feeling. We had significantly matured and finally felt old enough to be married. We accepted all the responsibility and the give and take of our relationship. Samantha brought us together and kept us together long enough to fall in love again. There was no question our hurried decision to get married three years earlier was the right one. We know for certain it had not been a horrible mistake.

We were personally happy, but Bill was professionally unhappy. Bored and frustrated, he looked at aerial photos of the USSR and Eastern Europe daily. The job he was doing differed from the job he expected. Mainly, he preferred the challenge of running a company of soldiers over being nobody's boss and being removed from the action. After six months, he came home every night either depressed, discouraged, or disgusted. Once a week, he talked to the person in charge of orders to see if he could get out of the job before his year was up. To make him more contented, I declared I wouldn't mind moving.

Thinking of our financial future, he once again talked of staying in the Army long enough to become a field grade officer, a major, or higher. He felt he needed a full tour in Vietnam to increase his chances. We believed the war would end soon since President Nixon was reducing troops and holding ongoing peace talks in Saigon. Returning to Vietnam as a military intelligence officer would allow him to be involved in rebuilding villages and schools, actually helping people. We hoped his time there would be more worthwhile and rewarding than just carrying weapons and trudging through rice paddies. We trusted in rumors that troop reductions meant many soldiers could return early.

He volunteered to be sent to Vietnam. I was not overjoyed at the idea. I accepted it as a part of the "military way." I found good points along with the bad, concentrating on the benefits it would bring— career-wise and future financial security. I had so many mixed emotions and wasn't sure which would triumph. I started thinking he wouldn't go this time or that his tour wouldn't last a year, but then I had to change my thinking as time drew nearer. Bill kept telling me that my cheerful acceptance and attitude toward the idea were helpful to him.

His orders to Vietnam came in July. When we shared the news with Jayne, her concern was about my feelings. We were most surprised by her offer to let Samantha and me move in with her, store our furniture, share expenses, and save money. She softened by saying that Bill and I had matured a lot in the last couple of years and that she wasn't under pressure now as she had been. She named all the reasons it would work, but most seemed to favor her wants rather than our needs.

I never considered her offer. I needed more space and was sure there would be a strain on all of us. When I wrote the news to my mother, I ensured she knew I wouldn't try to live with her and Dad either. I told her I liked doing things my way and at my own pace and sometimes being alone.

We were also ready to add to our family. We were excited about becoming new parents again, and I wanted to get pregnant, hoping for a boy. Merry-K, herself childless and unmarried, offered her non-supportive opinion that it would be emotionally challenging for Samantha if my attention shifted from her to a baby. She believed I would struggle emotionally without Bill's support.

I didn't argue with her; I silently reviewed my list of reasons:

1. Our life must go on.
2. I find an inner strength being pregnant.
3. We may not want to start over again if we wait longer.

We were determined to live each day as it came, kept trying to get pregnant, and hoped for the best.

CHAPTER 19
INTERLUDE
NOVEMBER-DECEMBER 1971

B Y SEPTEMBER, I had a new father-in-law. Jayne married a long-time family friend and former neighbor. His name was Lyle, and he'd come courting after his wife died. We gathered in the Covenant United Methodist Church in Springfield for the wedding. Although I had never taken Jayne's offer to stay with her seriously, it was now out of the question. Her happiness was also ours. She was delighted to have someone "take care of her" after she had taken care of others for so long.

Jayne prepared to sell her home and move with Holly, her youngest daughter, to Lyle's home in New Jersey. In the following year, they planned to move to Maine, and I knew we would not be getting back to Pennsylvania. We began our packing as well.

The sterile but sobering words on Bill's orders to Vietnam read: "MACV (Military Assistance Command Vietnam). Reporting Date 26 January 1972, Temporary Duty from 2 November 1971 to 16 December 1971 to attend CORDS (Civil Operations and Rural Development Support) Course Cl Nr 4-72 at US Army Institute of Military Assistance, Fort Bragg, NC 28307. Officer is a 2nd Tour RVN volunteer."

We only knew when he would leave and when he would return a year later. We didn't know where he would be located or precisely what his job would be. We made plans for me to go house-hunting in Spokane, looking for a place with space for our second child, now on the way. Bill would be at Fort Bragg, North Carolina, for a six-week course to become a Military Assistance Technical Advisor. His training would include an immersive language course taught by Vietnamese wives of soldiers, intended for him to learn enough of the language to recognize if his interpreter was lying.

Our household goods, including our washer and dryer, retrieved from Jayne's basement, were packed by the end of October. We drove to Fort

Bragg in our new two-door, gold Volvo Sedan. With a leather interior, a fuel-injected engine, and stereo AM/FM radio, it was comfortable, large enough for our growing family, and still economical on gas. We bought life insurance on the payments and expected to keep it!

Cousin Marla and her husband John were still stationed at Fort Bragg. We visited them there, sharing news of our mutual pregnancies. Samantha and I flew to Spokane on November 1. Similar to arriving on my mother-in-law's doorstep three years earlier, now it was my mom and dad who picked us up and took us home to the farm.

My days were filled with family visits, house hunting, writing letters to Bill, weekly long-distance phone conversations, and waiting for the mail. This time, his letters home were filled with details of exploring sites on his own, how his classes were going, and were more mundane and reassuring than any before.

For the first time, deciding where to live was *my* job. The house I found had two bedrooms, one bathroom, a living, a dining room, a kitchen furnished with appliances, hook-ups for a washer and dryer in the partially finished basement, and a downstairs recreation room. The yard was large enough for the puppy I must have known we would have.

An Airedale named Rags came in January after Bill convinced me I could handle both a puppy and a baby. What neither of us expected was that Rags would be a runner and able to jump the five-foot chain link fence I later needed to install.

It was a quiet neighborhood, with a pair of redheaded kids immediately across the street for Samantha to play with. A brother and sister-in-law of our party-line eavesdropper on the farm, Ed and Karen, lived three houses away around the corner. Karen watched other kids, was a reliable babysitter, and often updated me on local news from Edwall. I knew her information stream would go both ways.

I signed for our household goods in November, unpacked, and prepared the house for Bill's arrival. My parents cared for Samantha while I flew to Raleigh, North Carolina. Bill met me at the airport, and we had four days alone, the only ones we had since Samantha was born. The only ones we would ever have.

We drove west cross country, remembering our trip together when we had headed east three years earlier. As we discussed our separation in the year ahead, the outline of our dream with a perfect ending was clear. The baby would arrive in early May. Bill would return on paternity leave in late June. Half of his tour would be over. In October, he would be granted rest and recuperation (R&R), which we would spend in Hawaii. He would return to Vietnam for the remaining three months of his tour. He would be home in January.

We would delay our Christmas gift exchange until then. We imagined nothing but a bright future when his tour ended. We would have two kids (and eventually maybe more) and a dog (or two).

Bill said, "I sometimes think I should get out of the Army so we would never be away from each other again."

CHAPTER 20
BEING BRAVE
JANUARY-JULY 1972

W E HAD FIVE weeks together before he left for his second tour in Vietnam.

It was important to celebrate Christmas alone in our home. As we had in DC, we purchased a full-size tree from a lot instead of cutting one illegally off the military reservation. Except for the clothes I picked out, Samantha's presents had been bought by Bill while in North Carolina and were carried across the country in the trunk of our Volvo. A crib for her dolls, a chalkboard, and his presents for me.

We treasured every moment. A whirlwind of holidays and family events went by in a blur. Samantha held on to her Daddy whenever she could, stayed beside him, sat with him on the couch, and looked at him. We delighted in her precociousness—recognizing and naming letters of the alphabet on billboards and street signs. We prided ourselves on never talking down to her or using "baby talk." Bill expected her to speak correctly and often reminded her to say "Yes" instead of "Yeah." At least it wasn't "Yes, sir."

He left on January 21, 1972. Once again, I patiently waited for letters from Vietnam that would bring a usable address. As I had three years earlier, I recorded my voice on a cassette tape, saving the tape to be sent as soon as possible, and wrote down the activities of every day. I stayed busy, joined the YMCA, and signed up for bridge lessons and candle-making classes. I found an OB-GYN doctor and began to get acquainted with my neighbors. I spent time driving the forty-five miles to the farm for overnight visits and once managed to spin out on icy roads on my way back home, suffering no more than a flat tire.

Bill was assigned to Advisory Team #88 in Kien Hoa Province (later called Ben Tre) in the Mekong Delta, near where he had served his first tour. US troop levels in South Vietnam were approximately 133,000, down from the 1968–69 peak of about 549,000. As the number of US forces rapidly decreased, President Nixon planned to strengthen the Army of the Republic of Vietnam (ARVN) to make it capable of defending its country independently. Given the troop reduction and this policy known as Vietnamization, Bill was not assigned to a traditional US Army organization; instead, he was assigned to the joint-service Military Assistance Command, known as MACV. He worked alongside Navy, Army, and Air Force personnel and career civilian advisors.

His primary responsibility was to provide intelligence support. He gathered reports and information from various sources and independently analyzed the Viet Cong and North Vietnamese Army for strengths, weaknesses, activities, and objectives within the surrounding area. He also passed information on to his Delta Regional Assistance Command headquarters. He briefed his Vietnamese counterparts (through an interpreter) every morning on the latest enemy-initiated incidents, radio intercepts, and airstrikes. He oriented newly assigned team members on the topography of the Mekong Delta.

At the beginning of one tape, I heard his soft baritone voice describing the terrain, the lack of trees, the abundance of water, the dikes out in the open, the reliance on bridges and ferries for transportation, and the value of rice cultivated and grown in the Delta, critical to both North and South Vietnam.

The overall purpose of the teams of advisors, who were assigned and operating in each province, was to make the ARVN capable of independently defending its country. He was only the latest in a long string of advisors, sometimes barely tolerated by their Vietnamese counterparts or, at best, valued as equals. As a team, they acted as the point of contact for American support, providing helicopters for visual surveillance, medevac, and fire support. They advised what type of bombs, rockets, or other armament might be available for air strikes and how to use them with safety for friendly troops and civilians. They also provided financial support for development projects.

I wrote a letter or sent a card daily. Bill did his best to write every night, but sometimes, the heat, fatigue, or a duty assignment kept him from writing. His work sent him to Saigon once a month, and if he had time to wait in line for the phone, he called. He told me how much he appreciated every letter and kept them to read again. I not only dated the first page, I added the number of each letter I wrote. They often arrived out of order, and when he received several in a bunch, what I mentioned having planned for the following day in one letter would be reported on in the following letter he opened. Occasionally, one would go missing or get delayed. If he skipped writing a day or two, he reminded me, "I might not write as often, but I call more often."

The form of our letters seldom varied. We each began with an account of daily activities and minor frustrations, whom we saw, what we ate, and how we felt physically. We shared details about money spent, what we purchased, what checks we wrote, and if the checkbook was balanced. We kept each other up to date on movies we had seen or books we were reading or music we were listening to. We asked questions, knowing we wouldn't receive an answer for a week or more. The final page always included attestations of loneliness and declarations of love for one another.

With his camera in hand, Bill took pictures of the sights in the village of Ben Tre. He sent the film to be developed and the slides came to my address. It was helpful as I tried to visualize where he spent his time. The compound for Team 88 occupied a large city block next to the Ben Tre River and across a narrow side street from the Vietnamese province chief's official residential compound.

The advisory team compound was a crowded complex of many single-story buildings, the lower half built with white-washed cement blocks and the upper half heavy-screened and louvered windows, with sheet rock or tin roofing to shield rain. A small Vietnamese-style villa housed the small team club, mail room, and compound store. A high stucco wall topped with sharp, pointed steel arrows ran around the outside.

Bill was invited to the Province Senior Advisor's (PSA's) residence for dinner when he arrived. He was amazed by the new experience and wrote about what it was like *"to be in the 'combat zone' in civilian clothes, in a big old French house being served cocktails, with steak and potatoes and ice cream and pie and just sitting around bullshitting."* Such events happened every month as both a "hail and farewell" for those newly assigned to the area or those who were getting ready to leave the area.

He frequently shared his worries about financial support, wanting to be sure his pay was deposited in our bank on time. He set up his pay to include a monthly $25 US savings bond sent to me. When his salary increased to $1,156, he asked me to allot $156 to him for incidental expenses. He told me about every check he wrote and what it was for. I wanted him to have as many pleasures of home as he could manage, and he wanted me to be comfortable and secure and keep us out of debt. An essential minor purchase for him was an FM radio, hoping to hear classical music, the big band sounds of dance bands from our parents' era, and the love songs from movies we had seen together before he left. He also budgeted for three months, saving enough money to order a Mamiya RB-67 professional-style camera, which arrived in late July.

Often, he shared an amusing incident or details I didn't need to know about floor shows, which often began with performances by dancers and ended with naked women on stage interacting with GIs, but only after the Vietnamese counterparts had left the Officer's Club. Occasionally, he wrote words to reassure me he was behaving himself. He included a dialogue with two of the girls who worked in the compound who were sisters.

> One likes Captain Godby and the other Captain Delaney. Both were gone today. I said to the girls, 'Both of your boyfriends are gone, and you wait for them, right?' One said, 'Yes. You go to Can Tho and come back already, and no one wait for you. Why you come back?'

Most evenings, the four officers gathered, as Bill told me, "to tell each other lies." Liquor was cheap and tensions were high. A Monopoly game would be the premise. During their nightly get-togethers, his drink of choice was Seagram's VO Canadian Whiskey mixed with Vernors Ginger Ale (a nostalgic reminder of his visits with family in Michigan).

Relaxing together and telling jokes and tall tales happened in one person's room, where he informed me that "burning the blankets and spilling drinks on the bed" occurred. Bill commandeered the unoccupied spare room in his four-room billet, reminiscent of our hospitality in our apartment. Easter Sunday weekend, they emptied the space of a wall locker, a bed, and a foot locker, moved in a refrigerator, an old couch, a wicker chair with cushions, two folding chairs, a table, a floor lamp, and a bookcase. They made the space into a lounge. Bill painted the refrigerator. The doors of the three remaining rooms got coats of paint in three shades of green.

In one of his first letters, he wrote the following:

A little insight into Kien Hoa would have to include some bad news. Most of the people are Viet Cong Infrastructure (VCI) or communists. The woman who represents the National Liberation Front (NLF) at the Paris Peace Talks for North Vietnam is from this city. The second in charge of the North Vietnam women's army is from here too, and this was the staging area for a Viet Cong battalion during Tet '68.

It was as natural for him to tell me of dangerous situations he found himself in as it was for me to share news of my latest doctor's appointment. When enemy activities heated up in April, a week of his letters read like this:

Sat 8 April 2130 Hrs

Dearest Bonnie,

Kept busy this afternoon reading about all of Kien Hoa's enemy units, intentions, new weapons, etc. Not a bright picture. I ended up at the Thanh Tac Ferry site at about 1800 hours with the S-2, Police Chief, and Deputy Province chief. We've got info that they are supposed to blow up the ferry site again, destroy the hamlet office, and sink the boats off the landing the next few nights. Thanh Tác is on the road from Ben Tre, past the Truc Giang compound, north of the My Tho River. It is the last point in Kien Hoa on that road. Should be interesting to see what happens.

Mon 10 April 72 1700 Hrs

Dearest Bonnie,

I am tired. I am getting very tired of the same old routine. Work all day, bullshit at night, write a letter, have a drink, listen to the news, and then get mortared. I don't mind the first part, but this mortaring has got to quit. I'm getting just a little jumpy.

Last night, I was very rudely awakened by another mortar attack on this compound. Evidently, the Americans have pissed the VC off—3 episodes in 3 days. Last night, the first round landed across the street and wounded a soldier after going through the roof; the second round landed in the S-4 building inside the compound... 3rd round landed next to my bunker, damaged five jeeps/trucks, punctured radiators, destroyed carburetters and engine blocks, flattened tires and messed up windshields something fierce. 4th round landed in a jeep in the compound, blew the top out of the compound, and just absolutely wasted the jeep; it also blew out all the rest of the windows in all the rest of the vehicles inside the compound. What an unholy mess. The rest of the rounds landed just outside the compound, shearing trees and bushes to the ground. The bad part is that you never know when they are finished.

I have gotten over being apathetic about incoming mortars. Last night, after I finished the letter to you, I put my flack jacket and steel pot next to my bed, put two frag grenades on my desk, cleaned out under my bed, opened up a pair of boots in the middle of the room, put an extra pack of Camels and matches in my flak jacket and even set up the tape recorder to put down on tape the next attack. I am sorry I didn't get around to turning on the tape recorder.

I have become a light sleeper again, although a mortar round in the compound is enough to wake the dead. After the first round or rather during the first round, I rolled off the bed, grabbed the flak jacket, and was under the bed wrestling my flak jacket on in limited space in no time at all. After the 3-4 rounds and Dave Godby (another 6-foot-tall colleague) joining me, I got my pants and boots on while under the bed—quite a trick.

Well, lover, I think of you constantly and the safe, quiet nights in a real bed. Soon enough. I really enjoy your letters and tapes.

Take care, Kitten; I love and miss you.

Love ya Lots,

Bill

He sent me a map of Vietnam so I could follow his accounts. I didn't often do that. He encouraged me to subscribe to *The Army Times* to stay abreast of actions in Vietnam, but I knew it would create fear and uncertainty. Listening to reports on the evening news was challenging enough. I needed to hear directly from him a few days or weeks removed from the action. It helped me sleep.

Each month, he included a countdown.

26 February: *Well, one month down and eleven to go. For God's sake, don't worry about me; I'll worry about you.*

20 March: *By the time you get this, I will have been here two months—seems like yesterday, seems like forever.*

2 April: *3 months minus one week, and then who knows. I can hardly wait to get home, be near you, and sit with my arm around you.*

28 May: *One month from today, I'll be home, and if I had my druthers, I'd stay home.*

4 June: *Well, lover, three more weeks and three days, I'll be yours, not the Army's. Right now, that is the light at the end of the tunnel.*

When an audio tape arrived, I was thrilled to hear his voice, no matter if he spoke about mortar attacks and included recordings of their nearby explosions, if he asked mundane questions about the dog or our daughter or my doctor visits, or if there was money in the bank. I even played the tapes for family and friends so they would understand what Bill was experiencing and that there was still a war going on. I carefully listened for sweet, personal, and sometimes suggestive comments intended only for me, which usually came at the end of the recording. Those I kept to myself.

On Bill's recommendation, I added Carol King's recording of "So Far Away" from her new album *Tapestry* to my nightly playlist. I knew he had gotten himself an FM radio to listen to music in his room, and we kept each other informed on our new favorites.

Because Bill often remarked about how much Samantha must be growing and changing, I started recording tapes during the day. Samantha would inevitably come into the room with something to tell me. When

I told her I was talking to "Daddy Bill" and asked her to say something, she would be uncertain and sound shy or silly. If I didn't pause my recording a natural conversation about our day would be captured on tape. Years later, I learned that Bill played the latest tape each night, and one of his colleagues, Captain Brian Valiton, listened to Samantha's voice on the recordings as he fell asleep.

Our recorded interactions included his humor and gentle kidding, which made me miss him more. For example, in reaction to my letter telling him I missed him and now slept on his side of the bed, in his next tape, he said, "You've always slept on my side of the bed."

With my belly growing, I told him, "It now takes me five minutes to get my pregnant body out of bed." He reminded me that I have never been a morning person, preferring to roll over and grab another few minutes of sleep.

When I warned him that the wingback chair his mother had given us, for which I had painstakingly sewed a slipcover, was beyond repair, he replied, "No sweat about the passing away of our 'hand-me-down' chair. By next year at this time, I doubt that we will have your hand-me-down bedroom set. Then, when other couples tell of all the stuff given to them by their parents, we'll look a little sad and say we got nothing."

I laughed at all these remarks, cherishing the good-natured banter of our lives, even if there was a week-long gap between comments.

While I grew a baby inside me, visited the doctor regularly, set up a cradle at the end of my bed, washed baby clothes, and waited, Bill kept track of the due date, forgetting that Samantha had arrived a week late. On May 14, he wrote:

> *Well, I am waiting. What is it? A boy or a girl or just late? I hope you are not waiting for me to come home this time, too. Go ahead without me and tell the Red Cross, and I'll be home later... I miss you, lover, and I can honestly state that I will never volunteer for a hardship tour again. It isn't the danger or anything. Just missing you and the comforts of home.*

The warm weather of mid-May added misery to my swollen ankles and uncomfortable belly. I spent Mother's Day with my mother and my daughter, taking Samantha to a park that had a carousel she could ride. Mom assured me whenever I needed her, she could make it from the farm to Spokane in less than an hour. I just needed to call her. My cousin Cathy, whose wedding had been four weeks after mine, also lived in Spokane and had offered to watch Samantha when I went into labor.

Two days later, on Tuesday morning, May 16, I woke with the now familiar sense of contractions. I thought about my mother's promise but knew it was a day for her bi-monthly bridge club. I waited until noon, knowing she would leave for her bridge party in an hour. As I remembered my first false labor scare and embarrassment, I also realized that this time, I was alone. When my water broke, there was no time to stall any longer. No, Bill would not be there. Yes, I was ready. My parents arrived at 2:30 p.m.

We dropped Samantha at Cathy's, and I was prepped for delivery. My contractions were four minutes apart. This time, my labor progressed quickly rather than the prolonged eight hours with Samantha. I was given an epidural and episiotomy, numb from the waist down, but cooperating in the delivery.

Abigail Blythe Chandler entered the world at 6:58 p.m., weighing eight pounds six ounces and measuring twenty inches long. Mother was the surrogate father in the waiting room, and she was the first to know and get a peek at my daughter. She immediately contacted the Red Cross.

And then, the doctor couldn't stop my bleeding. I had gotten off birth control pills, influenced by questions about the long-term effect of the hormones on my body. We had relied on an IUD (intra-uterine device containing copper) for birth control in the year before I got pregnant. During that time, I had experienced cramping and heavy periods, which I was told were normal. I speculated that the placement of the IUD had scarred my uterine wall and that delivering an eight-pound baby had torn the lining.

In reality, postpartum hemorrhaging is a common complication. I was sedated, and the doctor performed a D&C, removing retained pieces of the placenta. Then I received a blood transfusion. While I recovered and my vital signs were being monitored, I was isolated from Abigail overnight.

It was a scary time. I needed to see her, feed her, and count her fingers and toes. I woke in the morning frantic to know if something was wrong with my baby. Once I asked, she was brought to me, and I fell in love all over again, crying as I clutched a surprisingly dark-brown-haired, hazel-eyed bundle.

I had an extra day to recover in the hospital. Our time coincided with Spokane's annual Lilac Festival. Royalty, high school seniors who rode on a float and led the torchlight parade on Thursday night and the Armed Forces Day Parade on Saturday, visited hospital patients. A picture of Bill was on my bedside table. A pair of princesses came into my room and swooned over Bill's photo, clearly moved by my story. I kept it to myself that it had been me on one of the floats in the same parades six years ago as a high school senior from a neighboring town.

Concerned by my description of the aftermath of Abigail's delivery and distracted from his duties, Bill got permission to go to Saigon to call me on May 26. I reassured him that the procedure had no residual effect. Yes, I could have more children. I was excited to hear his voice and included him in the stories of our two daughters, especially how Samantha was enchanted by her little sister. It made him feel much better, but we both wondered and wished, "What if there had been a true medical emergency? Maybe he could have been sent home and stayed home." Of course, in another era, or if I had delivered at home, I could have bled to death, and this story would have been entirely different.

At the end of every call, infrequent as they were, there was always something left unsaid. On the phone, Bill said he could not imagine being the father of two or having a baby in the house. His remark confused me. I certainly could not imagine NOT having a baby in the house. I had conceived, nurtured, and produced a baby with little more to think about over the previous nine months. She was living and breathing and taking my attention every moment of every day.

I decided to continue the conversation on the next tape I recorded. Sharing my thoughts out loud has always been the primary way I gain clarity. I told him I understood his doubt because I could not imagine his daily life. Then, to my surprise, ten days after giving birth, most likely energized by the blood transfusion I received in the hospital and influenced by the common statement of sympathy from family and friends—"I'm sure you wanted a boy"—I made an offer.

With hesitation and the sound of disbelief in my voice, I said, "If you are disappointed we didn't have a boy, I don't want you to tell me now. I don't believe I am saying this just a couple of weeks after Abigail was born, but maybe in a couple of years, if we could figure out the magical formula, we could try again for a boy. You'll be home just a month from now, and you might like this one. I think she is pretty sweet."

He was able to call again on June 8. I was trying to comfort a crying baby this time, but Bill hearing her cry helped make her a reality. I told Samantha her daddy was on the phone, and she asked, "Is he coming home?" It was a frequent question. She began asking in February after he missed her third birthday. I was happy to say, "Daddy will be home in three weeks." Soon, she told everyone she saw, "My Daddy Bill is coming home."

CHAPTER 21
ALL THE TIME IN THE WORLD

IN OUR CORRESPONDENCE, we finalized plans for Bill's paternity leave. He had fourteen days from 28 June to 11 July. Within this was travel time, giving us ten days to be together. The night he arrived, my parents waited with the girls at home while I picked him up from the airport. They stayed long enough to hug him and say hello, then discreetly left us alone. I had not thought about how the cradle with Abigail at the end of the bed might quench Bill's ardor. It took us another night to remedy the situation.

We spent many hours with my parents, Bill telling the stories behind the slides I had received. Viewing, sorting, and labeling the pictures made Vietnam a real place, the war real, and the people and places he wrote about came alive.

I knew my hometown friends would love to see him as they gathered for grilled steaks, fresh-picked corn on the cob, and apple pie on the Fourth of July, but Bill said he had seen enough fireworks. We made excuses and took the girls to my aunt's Bead Lake cabin in the Colville National Forest. We snuggled under the night sky as Samantha and Abigail slept, watching satellites passing through the Milky Way.

We looked forward to being alone in Hawaii. His twenty-fifth birthday was four months away. I vowed to lose more baby weight and be bikini-ready. He was disillusioned about the pointless war and wanted to leave the Army when his tour ended and finish his college degree. We discussed options of where we would settle down. East Coast near his mother? West Coast near my parents? On his flight home, over Portland, Oregon, he had seen the green pine trees and blue water and declared, "Portland looked nice."

We both decided there was no hurry, for when he came home, when he became a civilian when the war was over, we would have all the time in the world.

A week later, we stood arms around each other, waiting for him to board his plane back to Vietnam.

Bill said, "Be brave, Kitten,"

"I will," I promised. His hazel eyes and Cupid's-bow grin seared my heart.

"It won't be long now," he said. "I'll be home soon. Then we'll have time for us."

"It'll feel like forever before I see you again."

One last embrace. One long, final kiss.

I stood alone at the gate, scanned plane windows for his face, and watched, hoping to see the last wave of his hand. I held back tears as I walked out of the terminal and through the parking lot. I drove to the farm to retrieve our children, but seeing my tear-stained cheeks, Mother put me to bed. That afternoon, I cried myself to sleep.

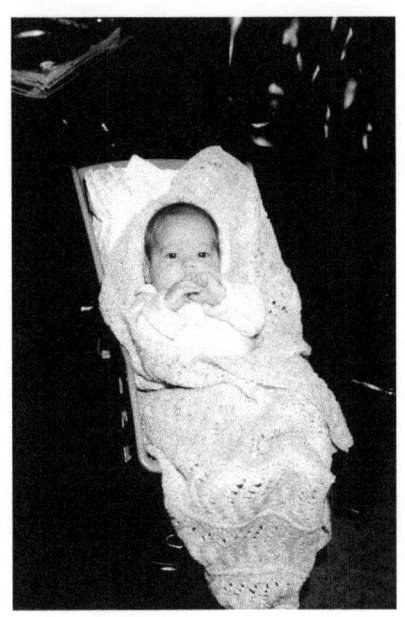

Abigail, Six weeks old, June 30, 1972
Picture taken by Daddy Bill

Samantha July 4, 1972
Captured by Daddy Bill

CHAPTER 22
"TAKE CARE OF YOURSELF, LOVER, AND KISS THE KIDS"

W HAT BILL WROTE in a letter dated 1 June 1972 could have been repeated daily:

Today brought no new hope of the war ending tomorrow, nor did it indicate it would last any longer than forever.

Shortly after Bill returned to Vietnam, he said our next assignment would be in West Germany for three years. It was the temporary solution to our dilemma whether he would stay in the Army or not. Our attention in our tapes and letters turned to making plans for the move. We saw ourselves together on an adventure. We were confident there would be untold opportunities for advancement and education. I began to look into ways to continue my education. We were excited that our clever, intelligent daughters would become bilingual.

Besides being busy with a newborn, I spent most of my time dreaming and planning about our time in Hawaii. I trusted I would keep losing weight and bought enough material for two new two-piece bathing suits. My thoughts returned to our conversations three and a half years earlier, the first time Bill went to Vietnam, planning another R&R. If he had not been wounded and sent home, our baby would have been five months old at the halfway point of Bill's first tour. Then I suggested I either bring the baby or fly to Spokane first to leave her with my mother. We never had a chance to settle that debate since he arrived home before Samantha was born.

This time, there was no doubt we wanted the time for the two of us. Alone. Together. I worked out a timeline for when I would need to wean

Abigail from breastfeeding and made an appointment to talk to my doctor about birth control. But first, I cut out the fabric for a new two-piece swimsuit and got out my sewing machine.

Bill had an overnight layover in Honolulu on his flight back to Vietnam. Put up in a hotel in an area catering to service personnel, he went out for drinks at the nearest bar. When it turned out to be a strip club, he decided I would not enjoy the neighborhood. We planned to spend R&R on Maui instead. I would arrive a day early and be there to meet his plane. I investigated prices on airline tickets and waited for a firm date in October. We needed things to look forward to that would see us through the long wait before we could begin the rest of our lives.

Bill called on my birthday, July 25, but I missed his call. I was on the farm with my folks, retrieving Rags after I had returned from a trip to the coast. I visited my sister and two college roommates and took our girls out for a day at the ocean. Two dozen red roses arrived a week later, and he called again from Saigon on Friday, August 4.

"Happy Birthday, Kitten. And Happy Anniversary."

"Oh, Bill, I can't believe you called. Where are you?"

"I got a ride to Saigon," he said.

"The roses came. They are beautiful. Thank you. I miss you so much."

"I miss you too. My visit seems like forever ago."

I drew in a deep breath and willed back the tears.

"I'll bring another dozen when I come home."

"Or a lei in Hawaii?" I asked.

He laughed, "A lay or a lei?"

"You know what I meant. I want you... to be with you..."

"Yes, I hear you." I heard the familiar exhale of cigarette smoke. "How are the girls?"

"I took them to the ocean. They loved it. I visited Janet and Dave, her new boyfriend, and Diane—I think you met her at our wedding. We spent the day at her parents' beach house. Her dad, Dr. Phillips, says hello. He was my advisor at UPS. I'll send you pictures soon. I'm sad we didn't take any pictures of you with the baby."

I suddenly realized I was talking to dead air.

"Bill, are you there? Bill?! Bill!" I cried out, "Damn it, damn it to hell."

I slammed down the receiver.

I waited, hoping he would call back. I sat watching the phone attached to the kitchen wall, willing it to ring. I opened a bottle of Annie Green Springs wine, the sweet taste reminding me of lovely nights together with Bill. I hadn't even said goodbye. I did not know it would be the last time I would hear his voice.

I took the second bottle and my tears to bed and listened to the rhythmic breathing of the baby. After Bill had left on July 11, I moved Abigail's cradle back from Samantha's room to the foot of the bed. He loved our daughters but wasn't prepared to make love with a baby two feet away. He was more comfortable with children when he could swing them in the air and make them giggle.

Meanwhile, the information Bill had shared in his letters in April was only the beginning of a worsening tactical situation in the province. B-52 strikes in Hanoi forced NVA (North Vietnamese Army) regiments into the Delta. His letters included mention of a colleague being seriously wounded on July 17, a few days after Bill returned. His colleague's Vietnamese counterpart was killed, and 140 other soldiers were injured. The district to the north, Ham Long, no longer had a DSA (District Senior Advisor) or American support. He mentioned the NVA moving towards Mo Cay, fifteen miles away, with outposts nearby being overrun.

23 July 2350

Mo Cay is being Mortared. I hope it stops there. There has been a considerable influx of troops, NVA included, into that area—they want the district town.

August 3

A rumor going around that a new NVA regiment just entered Kien Hoa (Province). Should prove interesting if true.

Take care of yourself, lover, and kiss the kids.

It was the last letter to arrive before I decided to take my little family out of the city for the weekend. Our tiny house grew warmer daily in August and held the heat all night. Abigail needed to be carried or rocked after every feeding, and only swaying with her in my arms soothed her. Her energetic chatterbox sister needed cousins to entertain her, so I packed a portable crib, diapers, dolls, the dog, and water toys on Saturday, knowing that in the middle of the wheat harvest, the family would gather for dinner on Sunday. I drove to the farm.

August 11, 1972, Ham Long District, South Vietnam
Picture taken the morning he was sent to Tan Loi by Senior Advisor, Mr. Kotzebue

CHAPTER 23
AN AUGUST DAY

I T IS SUNDAY morning, August 13, 1972.

I lie in the grass watching the sunlight flicker through the Lombardy poplars, whose fifty-foot height provides shade and a cool breeze as cotton-ball clouds drift by. On State Route 231, the north/south arterial between Interstate 90 and Highway 2, the owner of every car, pickup, truck, and slow-moving combine is a neighbor. I raise my head to watch a green sedan drive slowly past. I haven't lived on the farm for over six years, so I think it could be anyone, but it registers in some corner of my brain.

Mother comes home from church and is in the kitchen. Frying chicken sizzles as it browns in the pan. Dad is in the shed repairing equipment. Squeals of laughter come from the yard. My brother's children are giving my three-year-old red-haired daughter rides in a thirty-year-old little red wagon.

While nursing Abigail, I hear gravel crunching in the driveway. The green sedan with government plates had stopped at my rented home in the city. The next-door neighbor had sent them to talk to Ed and Karen, who might know where I was. Ed and Karen told the occupants of the green sedan that I was visiting my folks on the farm, but they were mistaken when providing my father's name. Dad's name was Gene Monk; they thought it was Gene Moos.

The car had driven slowly by, then three miles to town, driving through another three miles beyond to the farm of Gene and Marian. Word spread quickly. The party lines lit up with calls, asking, "Did you see? Have you heard? They were looking for Bonnie. What do you think it means?"

From the kitchen window, I see two uniformed soldiers—an officer and an enlisted man— speak to my father. I stand, swaying, kissing the baby's fine blond hair, waiting for them to come to the house, wishing they would disappear.

I had memorized the protocol for informing a survivor. The bravery required being prepared and rehearsing my part. I hope he is missing in action, or maybe only wounded, hospitalized again somewhere.

"The Secretary of the Army has asked me to express his deep regret..." I hear the words before they form into sound.

I complete the sentence. I clutch the baby tighter. They mention a time. He has been dead for two days. They talk of a combat mission. I hear that artillery rounds took his life.

All I can think is, "This wasn't supposed to happen." The words come out of my mouth in disbelief and anger. "He wasn't in combat. He went as an advisor. Tell me how this could happen!"

I swallow my sorrow, my tears. I am met with silence. This is the most challenging moment.

After a heartbeat and an intake of breath, the taller one, the Lieutenant Colonel, says, "I'm sorry. This is all we know. You will receive a letter with details from his commander. You will be assigned a casualty assistance officer. You can call him anytime."

I need to know. I ask, "When will he be sent home?"

"We don't know."

I desperately want none of this to be happening. I am sincere when I say, "This must be difficult. I'm sorry you had to tell me this."

My mother reaches for the baby. "Let me take her."

"No!" I hold her to my heart. "She's all I have."

Mother's face crumples with her grief.

Father takes hold of my shoulders. His embrace is clumsy but comforting. "I wish it had been me instead of Bill. I would take his place if I could."

"I need to put the baby down for her nap," is all I can say.

Upstairs, once she is sleeping, the tears begin. I fear I might drown or run dry.

I hear Bill's voice. "Be brave, Kitten."

"I will."

"It won't be long now," he says.

"Only forever."

One last kiss, indeed the last kiss. I touch my lips. If only I had held on tighter, longer, given words to my fears, begged him to stay.

I wake up and know that word has spread throughout the area. Aunt

Ruth arrives, and I meet her at the end of the sidewalk. I need a shoulder to cry on. I need to cry without explaining myself to my daughter. I need to cry without worrying about how hard all this is on my mother or my father.

Flowers arrive from my college roommate. I cannot breathe. I wish I had preserved the two dozen red roses Bill sent for my twenty-fourth birthday.

The neighbors come to call. They say how sorry they are, and then the conversation quickly turns to farming and the harvest. They swap WWII stories. That had been their war. It had been Bill's father's military service that inspired Bill's duty to the country. They all came home to the family farm, heroes. They were the fathers of my friends. Their wives mention the woman they knew who married the soldier who escorted her husband's body home. They mean to be hopeful, offering the possibility of another life, a fresh beginning.

I listen to the carefree chatter of Samantha telling Grandma, "My Daddy Bill will be home for Christmas."

I wasn't ready to tell her Daddy Bill would never come home. I wanted to wait, shield her, and avoid death's finality. I didn't have the words to explain it. I didn't have the words to understand it myself. Once again, I swallow my sorrow.

Samantha climbs into my lap. She repeats, "My Daddy Bill will be home for Christmas." She wants everyone to know.

I say goodnight in the following uncomfortable silence and thank everyone for coming. I gather my children.

I cradle Abigail until she falls asleep. Holding her against my heart, I want to dream of Bill, see his face, hear his voice.

In the night, I feel his presence. I reach out, crying. I wake, startled and alone. The tears fall again and again. I write a final letter to say that I will always love him, only him.

CHAPTER 24
FATHERLESS BABIES

G ROWING UP, THERE had been a radio on the kitchen table, tuned to the farm report every day at noon. When I was the last child at home, a small black and white TV took its place. Not only news at noon, but the evening news was consumed along with steak, baked potatoes, garden peas, and apple pie. Not much had changed between when I graduated high school and six years later when I came home with my two daughters that Sunday.

It was Wednesday night when KHQ evening news announced the death of a local man in the Vietnam Conflict. I was startled and reached to turn off the TV. Another surge of tears threatened to disrupt my well-shielded heart, to reveal the truth I was not able to share with three-and-a-half-year-old Samantha. It was proof I didn't need, that it was true, a real thing. I felt my life, my private grief made public.

On the day of the funeral, August 25, 1972, I was so tired of the part I had been forced to play that I would not allow that community meal everyone must have to get over and get on. I sent everyone away. Well-meaning, heartbroken relatives gathered up the road and Aunt Ruth's and Uncle Clem's. I needed to feel the solitude of being brave. I explained we had to catch a plane in the morning for the National Cemetery three hundred miles away.

In the morning, I dressed carefully for the day—a short pink and brown skirt with a see-through knit mesh top. An outfit Bill liked me to wear. After all, I was going to welcome my husband home.

The day began with an early flight. When we arrived in Portland, a full-size sedan was needed to transport eight of us to Willamette National Cemetery. The transport of his body had taken ten days. He had been flown from Vietnam to Andrews Air Force Base and then to Fort Ord, California, where he was cremated.

Again, my father gave up a day of harvest. Me and my daughters;

my mother; Jayne, my mother-in-law; Holly, my fifteen-year-old sister-in-law; and my sister Susan crowded in together. We awaited Bill's arrival, not at the airport but in the office of Willamette National Cemetery.

Others arrived: my sister Sandy; her three-week-old daughter, Diane; and four cousins living in Oregon. I have no memory of whether we walked or drove to his gravesite, but my mind is clear about the jolt of seeing his six-foot two-inch frame reduced to a shockingly small box of ashes.

Although our country church had been filled the day before, only a dozen of us were in the noontime sun at the gravesite. As in a nightmare without end, the honor guard was in place. Words were said. Taps were played. Three rounds of gunfire echoed over the hills and out to the river. The flag was folded with precision and handed to me.

"On behalf of the President of the United States, the United States Army, and a grateful nation, please accept this flag as a symbol of our appreciation for your loved one's honorable and faithful service."

At three months, Abigail still fit in her green plastic baby seat that had adjustable metal bars to change the angle. Most of the time, she was in my arms, her colicky stomach making her squirm and cry. When I held her, I felt our hearts beat together, the pulse of her father's life continuing in her bloodstream.

My bravery was on full display, still hiding tears from Samantha because I was not ready to answer her questions, gratefully distracted by the fussing of her little sister. I felt outside of myself, listening to taps, flinching as the gun salute was given, and watching as the flag was folded into a perfect triangle. I accepted the flag and the spoken tribute from a "grateful nation."

The young soldier had a second pre-folded flag, which he tried to present to Bill's mother. With all the indignation of her anger and grief, she stated. "I already have one! One is enough."

I swallowed hard, continuing to be brave. Bill's father had returned from WWII and died at forty-three from a heart attack. Of course she was angry. But our losses were not the same.

My hometown community had been caring for us for two weeks. After the graveside service, my cousins invited us to spend the rest of the day with them. We drove forty miles to Silverton, Oregon, where we were fed. I sat numbly listening to small children laughing and others talking around me, waiting to resume my life, to find a life, to pick up the pieces, to invent a future, to sleep without tears or nightmares.

We returned to the airport, sending Bill's mother and sister home to Maine. Before she boarded their plane, Jayne encouraged me to find love again, to marry someone else, to find someone to provide for her son's children, and to be happy.

I stood, incredulous at her suggestion, still dressed in my provocative outfit, holding Abigail's baby seat, now filled with a folded flag. I felt wondering eyes upon our sadness. I wanted them to look. I wanted to be seen. I needed the world to know what it had done to me, to us, as they gazed upon a twenty-four-year-old widow with fatherless babies.

CHAPTER 25

GRIEF THAT NEVER ENDS

"What is left is a new you, a different you, one who will never be the same again or see the world as you once did. A terrible loss of innocence has occurred, only to be replaced with vulnerability, sadness, and a new reality where something like this can happen to you and has happened."

Elisabeth Kubler-Ross

I READ AND reread condolence letters. I look for an answer.

Letters came from President Richard Nixon; Washington State Governor Daniel Evans; The Secretary of the Army Robert F. Froehlke; the Commanding Officer of the United States Military Assistance Command, General Fred C. Weyand; and The Acting Chief of Staff of the Army, General Bruce Palmer. Jr. All were men who never knew Bill. All had signed the letter, typed by someone else, and placed before them. These were just a few of the letters they signed. They would not be the last.

Deputy Province Senior Advisor. Lieutenant Colonel William Tausch and Province Senior Advisor Albert Kotzebue dictated more personal words, words of sympathy that I believed and cherished yet did not fully appreciate for years. Their loss was also personal.

Mr. Kotzebue's letter described the day's events, but I could not absorb the details of what it said, still asking how and why.

11 Sept 1972

Dear Mrs. Chandler,

I extend my deepest sympathy to you in the recent loss of your husband, Capt. William G. Chandler, in Ham Long District, Kien Hoa Province, Republic of South Vietnam, on 11 August 1972.

On that day, a large enemy force had attacked Tan Loi village, some five miles from the Ham Long District Headquarters, and the village was in imminent danger of being overrun. A relief force of two battalions was sent in the late afternoon, with Bill going along to advise and assist the District Chief. The command group, where Bill was located, was proceeding along the road when they came under heavy enemy small arms, mortar, and rocket fire. Bill was killed instantly by a high explosive shell at approximately 5.00 PM. Memorial services were conducted at 3:30 PM on the thirteenth of August by Chaplain David O. Golden, the Protestant Chaplain serving our units in the Team 88 Compound in Ben Tre.

As a team member, Bill was liked by all his associates. He was an excellent officer who performed all his duties in a cheerful and efficient manner. His unfailing good humor and dedication inspired all who knew him. We are all saddened by his passing; we are all also most proud of his loyalty and devotion to his country's protection.

Bill's personal effects have been collected and are being sent to you. I know that you will treasure his personal possessions, and I hope they reach you without delay.

The sincere sympathy of all members of Team 88 is extended to you in your bereavement.

Sincerely,

A.L. Kotzebue

Province Senior Advisor

I got out the map of Vietnam Bill had sent me and traced my finger on a line from Ben Tre, where I knew he was safe in his compound, to Tan Loi. It made no sense. He told me about traveling throughout his district, but this village seemed to be in Ham Long, another district altogether.

When Bill's personal effects arrived at the farm, there was a stark moment of accepting he really was dead. His new camera, which had been an important step towards his desire to become a professional photographer; his FM radio, which brought him comfort, listening to music we had both loved; his cassette tape recorder, which provided a means of speaking to me and of listening to my voice; pictures he had taken of me the summer before, scantily clad, which he kept beside his bed. All of these would never have been left behind by choice. I had hoped the letters I sent to him might be in the green duffel bag, then remembered his telling me how the thin paper I used did not survive in the high humidity, and how the variety of stationery could not be neatly kept.

Samantha and I made the front page of the Spokane Daily Chronicle on February 7, 1973, receiving the posthumous awards of the Purple Heart and the Bronze Star for Valor in Bill's name. The citation was read:

> For heroism in connection with military operations against a hostile force, Captain Chandler distinguished himself by heroic action on 11 August 1972 while serving as a senior advisor to a Vietnamese Regional Force Task Force in Ham Long District, Kien Hoa Province, Republic of Vietnam.
>
> Captain Chandler, accompanying a two-battalion force in Quoi Thanh Village, was moving with the headquarters element when all elements, in a short interval of time, came under intense small arms, automatic weapons, mortar, and rocket fire. Unconcerned for his own personal safety, Captain Chandler rapidly assessed the tactical situation and reported the location of the enemy forces. He was in the process of directing tactical airstrikes on the enemy positions when his counterpart ordered the headquarters group to move to a more secure area in an outpost to the south. After waiting for the heaviest fire to subside, the group began its retrograde movement. It was at this point that Captain Chandler was struck by a B-40 rocket round.
>
> Information gained from his observations on the ground enabled the Sector Commander to place tactical air and gunship strikes on the enemy's positions, inflicting heavy casualties. Captain Chandler's heroic actions were in keeping with the highest traditions of the United States Army and reflect great credit upon himself and the military service.

At the time, the ceremony was a symbolic end to the details of Bill's death. I temporarily let go of needing more answers. The questions were no longer evident. I tucked grief away the best I could. I was swallowing my sorrow in order to move forward.

The words I cherished most came from the heart of a stranger.

August 18, 1972

Dear Mrs. Chandler:

Please accept my most sincere sympathy on the loss of your husband. Also, as a citizen, I would like to offer my gratitude for his sacrifice and to express pride in him for serving his country.

I do understand how you feel, for our son was also killed in Vietnam. I wish there were some way I could truly help you, but I can at least assure you that the horrible sick feeling inside you will gradually become less acute.

Sincerely,

Robert G. Heskett, M. D.

Chasing away that horrible, sick feeling became my goal. I faced the following year with advice from others, some of it coming from the feminist movement. I bought a new poster for my kitchen wall declaring, "A woman needs a man like a fish needs a bicycle." I was determined to return to school, earn a degree, begin a career, and support myself and my daughters.

The US government sent monthly payments. They were "survivor's benefits" and I resented each one, telling anyone who listened, "There isn't enough money in the world to replace a life cut short, a future destroyed, and my love taken away."

In truth, the money could have been enough, yet I never wanted to have the sole responsibility for my children and lack the companionship of another adult to share my life.

I turned to the wisdom of others forced to walk the path before me. Some advice came from a book called *Widow* by Lynne Caine. I took to heart the author's suggestion to make no permanent decisions in the first year. The exception was that within three months, I purchased a home in Spokane and took advantage of the final relocation move the military offered. Also, within six months I started college classes seeking a sense of purpose, planning for a career in education, seeing a future with me and my girls alone.

In November, I traveled with my parents to England and Turkey. My brother-in-law, Dick, was stationed in Izmir, as an Air Force Liaison with the Sixth Allied Tactical Air Force. Mom and Dad thought they should cancel the trip, not wanting to leave me home alone. I volunteered to go with them, taking the girls along.

We were in London for one week, riding double-decker buses, going to the theatre, and sightseeing. I insisted on going to trendy Carnaby Street and buying myself a full-length, brightly flowered, orange, yellow, and gray flounce-sleeved maxi dress. It hung in my closet for years, but I never wore it.

We spent three weeks in Turkey, including Thanksgiving. Surrounded by the buzz and activity of three noisy cousins under ten, hearing conversations between adults, and the street noises outside the apartment, I often nodded off. This replaced my loneliness and temporarily stilled my anxiety about what lay ahead for us.

I spent a day touring with an English Army Colonel who allowed me to tell my tale, express my loneliness and desires, and listen to his sound advice "to be strong, but flexible—to give my most to others and to wait for happiness."

On the transatlantic flight returning home, my family and I exchanged seats in the rear of the plane, where Dad had been able to smoke, for seats in the front of the plane. The front of the plane had a bassinet attached to the bulkhead where I could lie a fussy baby. When Dad walked to the back of the plane for a cigarette, I was unaware that he had engaged in a conversation with a stranger. Suddenly, a tall young man came to offer to walk with Abigail up and down the aircraft aisle. Of course, Dad had told him my tragic story and may have even been matchmaking.

Rags, now a year old, was cared for by Aunt Ruth and Uncle Clem while I was away. He had been an irritation and handful for them. I needed to decide if I wanted to bring him into the house I had purchased by using a down payment only possible because of Bill's life insurance payout. His pistol and his camera were among the items returned from Vietnam. Though they were important to him, they served as reminders which made me miss him more. Without him to raise and train a dog, I had to admit I wasn't in a position to raise a puppy. Rags found a new home with another aunt and uncle with acres to run and spent his days recovering from frequent run-ins with the local porcupine.

Back in Spokane, I wanted to create a place of hospitality, inviting family, parents, aunts and uncles, cousins, and their children to my new home. I hosted an open house in February and again on the day of Abigail's baptism.

I rejoiced in being surrounded by love and support, and I rediscovered that I enjoy cooking and entertaining and could serve as host and hostess by myself. I accepted that the financial security Bill always wanted for us was realized by his death. The purchase of a home, the ability to invest in stocks and bonds, and even having the Volvo paid for by insurance were reminders of Bill's maturity and responsibility. He managed his life to provide for our happiness and security, making provisions even in his death.

On Friday, January 26, 1973, I was acutely aware it was the date Bill would have completed his tour and come home. In a diary entry, I wrote:

I've tried not to think about it, but it's been on my mind all week—all of the last five and one-half months—all year. In a way, I'm hoping that, having passed this date, I can go ahead with a new life. Maybe my system can now finally accept that Bill won't be back. Maybe now I'll begin thinking of him less, but I can't see that. I'm living now on the memories of good times, but realizing there are no more good times makes me feel frightened and helpless. But the loneliness is worse.

On Saturday, January 27, there was an official signing of The Paris Peace Accords, which was signed after four years of negotiations. The direct combat role of the United States in the Vietnam War ended that day. It meant that all troops would be withdrawn and POWs would be released and brought home. Once again, the realization that Bill died in that war came back just as complex and challenging. I lied to myself, wanting to believe Bill's death may have contributed to the end of the conflict.

To soothe my inconsolable tears, Samantha told me, "Don't worry, Mommy, we can get a new daddy."

It was never that simple, but I tried. Friends began to set me up on dates with recently divorced men, men who had no desire to raise another man's children, and men who had never served in Vietnam. I invited some of these men to the farm for Sunday dinner, rehearsing and reliving the significant moment of "meet my parents."

Aunt Ruth knew about these Sunday visitors and said to Mother, "Bonnie shouldn't be in any hurry to marry again because there will be many men who want to marry her."

Between taking classes at Eastern Washington State College, studying, going on dates that involved copious amounts of alcohol, and raising children, I was tired all the time. I listened to frequent lectures from Dad about not drinking so much and from Mom about getting more sleep. I knew they wanted to see me happy, but I felt they tended to discourage pleasure.

I was numbing the pain of loss in a variety of ways. Anything that made me feel alive again. Alcohol helped. So did sex. However, letting an evening end with a new man in my bed and having one of my daughters wake up with a nightmare or a stomachache, I realized that I remained the only one responsible for their wellbeing.

CHAPTER 26
DAUGHTERS' DEVOTION

ALTHOUGH ONLY FOUR, going on five, Samantha became my companion and confidante. My parents' grief was as palpable as mine and tinged with the helplessness of not knowing how to ease it. Mother remembered Bill's honest disillusionment with the meaning of our presence in Vietnam and surmised that Bill may have come home a bitter man.

In our conversations, I agreed, though my knowledge was tempered by our shared tenderness and determination for a brighter future. On a visit to the girls' pediatrician, I passed out and slumped to the floor from exhaustion. My fatigue and grief overwhelmed me, and when we got home, I sat on the coffee table, unable to stop my tears. Still unsure how to speak to our daughter about why Daddy Bill was not coming home, I told Samantha how difficult it would have been if her Daddy was angry or somehow different. To her immature mind, that meant two things. One, she thought I was glad he was not coming back. She couldn't understand why he did not love her enough to stay home. These unspoken misunderstandings persisted for years.

As Jayne had encouraged me, I remarried quickly. I pursued my vision of being married to a responsible man who would provide financial security, love me and my daughters, and, because he was a veteran of the Vietnam War, understand my grief. We had a daughter together, acquired a dog, and I learned to laugh again. I also faced a new reality that not all men were as responsible, and none could replace my first love, a war hero.

Whether it was my moving on to a new life or because of her grief, we seldom heard from Jayne. Samantha was eleven, and Abigail was eight when my second marriage failed. Samantha's only memories of her father came from photos of the two of them together; Abigail had none. Amid another transition, it was the best time to go to Maine to visit Jayne and her husband, Lyle.

Bill's sister, their Aunt Holly, had completed college and was engaged to be married. It would be my last visit to Jayne and her family by choice and circumstance. We combined the trip with time with my sister Susan and family, now in Rhode Island.. Our visit was brief.

Jayne invited Samantha to return for a more extended visit. Two years later, Samantha flew to Maine alone and stayed for a month. She became better acquainted with Aunt Holly and assisted Jayne in her antique business. She was embraced and entertained and experienced moments of criticism similar to what had caused the original rift between Jayne and me.

The year Samantha turned eighteen, she wrote a letter to me on my birthday.

July 25, 1987

When you were my age, could you have possibly imagined your present life?

At the age of 18, you were planning for your first year of college with dubious anticipation. You saw for yourself new adventures that would lead you to a safe life filled with normalcy and happiness. A wide-eyed girl who had, perhaps, a clear vision of the husband and family her life would revolve around after college. How could you ever imagine that in six years, you would experience your greatest elation and your most dreaded horror?

Possibly, and I believe it to be so, I have experienced more in my first 18 years than you had. Therefore, I am not as naive or idealistic, although I realize I still possess those qualities somewhere in the recess of my being.

I believe that because you have allowed me to explore life as I need to, you have allowed me to get beyond the petty concerns of many conservative families. I know that money, being beautiful, having a wonderful husband/boyfriend, and owning many things can never be that important to me because you will love me no matter how I look, dress, live, and feel, no matter who I love or live with. I can be my own person, unconventional as that may be.

Maybe this is all owing to the fact that when you were in your 20s, every conventional thing you depended on had been blown apart by a war that we couldn't understand. Then again, in your 30s, family life was turned helter-skelter, and you had no choice but to change from what you once saw as conventional to unconventional.

I know so much more at my 18 than you did at yours, but I can still not imagine going through what you did by the age of 24. And simply because you already experienced those things, I feel I won't have to.

Because you have continued to love and work for that love in a world that has shown us so much hate, I can also see what good the world may offer me.

Thank you for sharing your suffering with me so I might grow.

Thank you for sharing your joy with me so that I may be happy, also.

Thank you for loving me and allowing me to love the world and its people without giving up hope that someday there will be no more war

I love you dearly, Sam.

She wrote the following essay the same year, her first year at Evergreen State College in Olympia. The assignment was to describe a dream or a vision:

The Hostage

I was walking alone one starlit evening toward the edge of my uncle's farm. The breeze remained warm after a sweltering day of harvesting. I had been driving truck all day and was still coated by the dust that clung to my sweaty face. I had to wait my turn to use the bathtub, so I walked out in the disced fields, waiting for next year's crop.

The silence was comfortable and safe, unlike the nights I had spent in the city. I had no fear of anything; the vastness of the open field sheltered me. There was no one, nothing as far as I could see.

I felt, however, a particular melancholy on this day because it was August 11, exactly fifteen years after my father had been killed in Vietnam. I sat down on the cool brown earth and let my soul fill that night with grief and sadness that I often think about but

---

seldom let overwhelm me. As my tears plowed through the dirt on my face, they dropped to the ground, coloring it muddy and dark.

In my mind's eye, I could see my father. At first as a father and husband and then as a soldier. The longer I cried, the more I could see of the war. I could imagine him beside his fellow soldiers, drinking and smoking cigarettes, talking about their families and life at home. I wanted to know those men. To hear their tales and learn more about my father.

Soon, I imagined my father's grave among a multitude of military headstones. People who died in battle or at home but had all served their country in a manner they thought was right. I saw the Vietnam Memorial and again felt sorrow for all the victims of war.

As I gained a clearer vision, I saw my father's name on the wall. William G. Chandler. But when I glanced beside it, I was shocked at the unexpected name I found. My mother's. I was filled with horror. But my mother's not dead; what does this mean? I was scared to look further, but morbid curiosity forced me to. As I finally allowed myself to direct my gaze, I noticed my grandmother's name, my aunts and uncles, my sister, and ultimately myself. I was frightened and confused. I considered the possibility of all these people being dead, but my existence was enough to prove to me that it couldn't be true.

This was my own memorial. I began to understand that what is represented was that all of these people had been touched by war. A part of each had died along with my father. His death robbed us of much happiness and filled us with inexpressible sorrow. The names after those of my family were mainly unfamiliar, but the ones I recognized showed me that these were friends of my father. Some from high school, some from the military. All people who had survived the war.

And then there were Vietnamese names. Local villagers who had been relocated by the US military only to be killed by the Viet Cong along with everyone else in the compound. Little children my father had photographed. People I had seen pictures of but knew nothing about, except that they lived in a war.

When the Viet Cong would invade, they weren't aiming for "Luc Dhong" and his family or Captain William G. Chandler. They were fighting the enemy. They were after the US military, not a 24-year-old father of two. I know that my father wasn't in Vietnam because he wanted to kill someone he didn't know; he was working to defend his country in the way he thought was right. But after six months in Vietnam, he wasn't even sure anymore that his presence there was what the Vietnamese needed. But since he was a loyal soldier, he was an enemy.

After I had exhausted my supply of tears, I wiped my eyes and prepared to head back to the house. My grandmother was aware of what day it was, and I knew she wouldn't question my sadness. As I walked back, my sorrow changed to anger. Anger at how war is such a non-personal issue. Enemy soldiers are nothing more than targets and obstacles for democracy to tackle. The threat of communist overthrow propagates paranoia, and we allow ourselves to ignore what's going on around us. We become the blind victims of another "ism."

As I went inside the house, I headed straight for the bath. As I undressed, I thought that somehow, somewhere, people must become aware that war is nothing but a personal issue and that we must each do our best to prevent history from repeating itself. We must allow ourselves to see those we are fighting as humans and to think of those who genuinely suffer from American attacks on foreign villages.

I settled down in the hot water and let it burn my skin. I washed the dirt from my filthy body and thought of a woman my age in Vietnam doing the same thing. Perhaps another child of a dead soldier.

No one meant for the Vietnam War to go on as long as it did. No one meant for the US or Viet Cong to kill innocent villagers. No one meant to kill my father, but he's dead.

Samantha's grieving continued to seek solace and find a voice. In the spring of 1988, she traveled to DC with her college roommate, Kimiko, and saw the Vietnam War Memorial for the first time. There, she met up with her Aunts Cherie and Holly and traveled to North Carolina, where her Uncle Chris joined them at Grandma Jayne's house. On returning to school, she made a narrated video recording with pictures of her father, entitled "One of the Fifty Thousand." It was an assignment in her Media Class in the early days of technology. She put voice to her pain and anger, expressed the sorrow and loss in her life, and exposed the cost of war, not only to us but also to other families of all those names commemorated on the Vietnam War Memorial, The Wall.

She carried a VHS copy of her video with her as she gave herself a six-month break from school and traveled to Louisiana, North Carolina, and Maine, spending a month with each one of her father's siblings, spending time around them—some of whom were big and boisterous and some that were single having recently gone through divorces. She took temporary jobs to be independent and had the chance to get to know people who had been virtual strangers, reporting that loud music was always playing in every household. Her quest was to understand the people she came from, and she needed to find ways to discover who her father may have been.

After college, Samantha pursued her interest in the world of acting. As the oldest child in our family, she directed her sisters and four or more cousins in backyard productions of well-known fairy tales. In her early teens, she began acting in Children's Theater; as a young adult, she found roles in community theater. Always wanting to improve her craft, in 1999, she attended the fifteen-week summer semester at the American Conservatory Theater in San Francisco. Her letter below was addressed to me at our cabin at Bead Lake—the place we had gone with her father on his final visit home; the place Bill and I had envisioned our future with a happy ending, the place I was able to purchase from Aunt Lois with the insurance money from his death.

August 11, 1999

Dear Mom,

The spinning of the world and the journey of life continue to amaze me. We are in the final week of classes here and are watching the performances of all the other groups throughout the week. Mostly, we know nothing about them until they begin.

The first performance we saw today was entitled "Reflections of War," and one of the final scenes was of a woman finding out her son has been killed in Vietnam. The story, of course, takes me back to your story of hearing the news on the farm, the car coming up the drive, your pity for the soldiers, Etc.

By the end of the performance, I was certainly in a daze and needed to be alone. I wanted nothing more than to be with you and my sisters and to share in our love and loss, and it was as I was walking down the street that I realized the date.

August 11

And the perfect eeriness of it came and surrounded me. I wanted to call but didn't have my wallet, so I couldn't. Instead, I thought I would write and tell you I love and respect you immensely.

This is a sympathy card 27 years late. A thank you note for all of the strength that you have shown and given to me. It is a wish that you continue to find joy and happiness, and aliveness amongst the bitterness of this world. You are strong and powerful—and with each passing year, I find a greater appreciation for all that you have survived and all that you have accomplished.

I love you.

I am sorry for your loss. Thank you. Samantha

In her early teens, I overheard Abigail telling some friends, "I never knew my dad. He died when I was a baby." They comforted her, letting her cry, and since I was eavesdropping, I did not know how to address her sorrow. It echoed mine.

Years later, Abigail also traveled to Texas, North Carolina, New Hampshire, and Maine to visit Bill's sister Cherie, his brother Chris, his sister Holly, and their grandmother. Each daughter grew in empathy with me, sharing life and my journey. Abigail and Samantha both retain their father's surname.

Abigail has honored and remembered her father by naming her sons after him. Her oldest son, born June 27, 1999, was initially given the first name Chandler. When she divorced and restored her maiden name,

Abigail decided Chandler Chandler would be too much, so Tristan became his first name. Her younger son, born ten years later, is a namesake, William G. Chandler, delivered May 14, 2009, two days before Abigail's thirty-seventh birthday.

CHAPTER 27

A WIDOW'S LAMENT

I N THE SUMMER of 2009, I was studying for a Doctor of Ministry degree. My career as a pastor was prompted by understanding myself as a "wounded healer" as described by Henri Nouwen, with the advantage of knowing grief and loss firsthand. Yet, still celebrating the birth of a grandson, recalling the death of his grandfather, I found myself writing a lament in the format of the Psalms, needing to express my sorrow and anger, wanting to offer thanksgiving.

June 21, 2009

San Francisco Theological Seminary

Trauma, loss, and grief in theological and pastoral perspectives

Bonnie Chandler-Warren

A Lament

Wiliam G. Chandler November 18, 1947–August 11, 1972
Sweet William, William's son, dearest love
Off to war, lost to death
"We regret to inform..."
I want to die.

Pain filled heart broken disbelieving
Letters sent but never read
Letters come and I feel your breath
Is there a thin place we can meet?

Your death is out of order.
My father would take your place if only you would return
Neighbors stay too long, not knowing what to say
To one so young

Our baby cries
Holding rocking feeding
I stay connected to your life within me
Sleeping wet with tears, I sense you near
Startled awake, your spirit vanishes
O God my God, where has love gone?
Will there be a time for its return?

In days of nightmare slowness
War is ending,
Fewer deaths
Yours but one

Your mother comes, and I am brave
It is what you asked
That final day of tender kisses and waved goodbyes
I hold my tears so no one can see my deep despair
The emptiness of my arms

Waiting for the plane with precious casket cargo takes so long
I wait I watch I pray
Take this cup of grief
Until the day there is a flag-draped box before me

Our baby cries
I rock and hold her to my heart
Bugles play rifles fire flag is folded
And presented "On behalf of a grateful nation..."
I want to die to scream to run away.

Still, we hear the drums of war
Others know the pain, the loss, the grief, the truth
No distinctions in gender, age, or nation
Children lose their fathers/mothers
Women/men lose their lovers
Parents lose their children, and we cry
Loss upon loss, death touches all our lives
No one seems to know or understand
Or speak a word for peace

In resurrection style you visit only once
My disbelieving father stops
while working in a field
Knowing you are near
He hears your voice,
"Tell Bonnie not to cry. I'm okay. I love her still."
He believes and knows God's tears.

In that thin place where you have come
Reaching out to bring me peace
Giving faith to one unsure
For comfort, joy, release, give thanks

For thirty-seven Augusts, I remember
Wonder what would have become of us
Our love, our life?
Always you are in our daughters' smiles, our grandsons' eyes.

I hear your laughter, know your voice
Reaching through that thin place
In Prince William, our daughter's son
Namesake, precious life, gift of God
I hold you to my heart again
Love reborn hope renewed life returned

May peace prevail on earth
Now and evermore
Amen and Amen.

CHAPTER 28

BRAVERY REVISITED

I T IS SATURDAY, May 25, 2019. I step outside a noisy, crowded Philadelphia restaurant to answer my cell phone. I have permitted the local reporter from the Springfield paper to interview me, looking for the human-interest angle on a local man who died in Vietnam in 1972. Bill was raised in Springfield, and his children and I lived in Spokane for ten years before relocating to the greater Puget Sound area in 1983. Since no family members live in Pennsylvania any longer, we had been hard to locate. Bill was the last Vietnam casualty from their town to be honored.

The reporter asks me for details, those annoying questions that now come within the first forty-eight hours when news crews and cameras are dispatched in every crisis or tragedy. She has read the comprehensive narrative posted on the American Legion website. It includes excerpts from his letters home and how he questioned the reasons for this war. She also read his poetry. She has seen my statements about becoming a pacifist, abhorring the cost of lives lost. She starts with softball questions.

She asks, "How did you manage to raise your daughters?"

"Well, I had my parents and the community I was raised in. I also didn't have options. I did what I had to do. Bill expected me to be brave. I practiced being brave."

"Do you remember how you were notified? Can you tell me about that?"

"I have never forgotten. I was home with my parents for the weekend. Every soldier's wife needs to be prepared for the two men in uniform to arrive at her door with bad news, to live expecting it. I wasn't prepared. I didn't expect it. I relive it every summer. I don't go to the farm without thinking about it. We didn't believe it could happen since Nixon was withdrawing troops. I remember asking, 'How could this happen? He was only an advisor. He wasn't in combat. This wasn't supposed to happen.' But it did."

I am feeling agitated. I don't want to answer these questions. I want to rail against this obsequious intrusion into my grief, the grief I have lived with my entire adult life. Yet, I remember my pledge to Bill to be brave and my decision to come accept this honor from this place. I want to be non-political, to be gracious, and perhaps find solace.

"How did Bill manage to serve in the military while he was a pacifist?" she asks.

"Bill wasn't a pacifist. But he didn't think we were in the right kind of war and felt that what he was doing wasn't helping anyone. It had gone on too long. I was the one who *became* a pacifist because of what happened."

We volley back and forth for another fifteen minutes, me offering too many details, wanting her to hear my pain and anger, understand it, and represent my life and views.

She ends the interview by asking, "Can you tell me what this means?"

I miss the direct simplicity of this question. She may be asking how it feels that he was being belatedly remembered and honored for his sacrifice. She may have been asking me to say something about returning to Pennsylvania after so many years. All I think of was how the years have passed without knowing the meaning of his death, how a twenty-four-year-old girl was unprepared for widowhood. How raising two small children alone is no easy task. Having imagined and planned for a life like my mother's—with one lifelong love to provide home and security—and that Bill's death had set me adrift.

I mutter something along those thoughts.

We have time on Sunday to explore Philadelphia. I make sure we make it to the Reading Terminal Market where everyone can enjoy an authentic Philly cheesesteak. I remember the moment with Bill and recognize my attraction to all other public markets.

On Sunday night, Holly, Bill's younger sister, and her husband came to the Bed and Breakfast my family and I had rented for the weekend. We had seen each other only once in the years since his funeral. Now in her early sixties, her facial resemblance to her mother caught me off guard, and I turned quickly away after telling her, "I saw your mother's face." Holly was stung by my reaction.

In 1968, she was only ten. Her brother was her hero. She admitted that she didn't understand the dynamics then and thought of me only as the girl whom her brother had adored. She knew my relationship with her mother was complicated, but she did not know the reasons why. I guessed she had not known her mother's suspicion that I had trapped her son into marriage with my pregnancy. Holly had also been shielded from the confrontation over money and Jayne's accusations of our irresponsibility that led us to move out of their house unexpectedly. I was curious to know what she remembered but decided to let the conversation drop.

We are seated around the table for dinner. Always the one to bravely speak through tears, Abigail laments, "This is so strange for me. I have heard only one story about my father. Not how or why he died, but how he met my mother. He will always be twenty-four, the age he died. I expect those who knew him to be frozen in time. It isn't right that they are in their seventies."

This time, I am the one stung by words.

Yes, I know. Four and a half decades ago, I chose to end the chapter, close the book, start a new life, and pretend. It was easier for me to bury memories than to bury a husband. If I didn't talk about him, I could live without him. I didn't do either very well.

If I married another man, I couldn't admit my love for another. I became an expert at compartmentalizing my feelings. I only cried when no one was watching. Letters and audio tapes remained safely in a wooden box that held the black and white photos Bill had taken, printed, and mounted. Carousels with slides taken on our weekend drives around the Washington, DC, area sat in storage alongside those filled with his slides of Vietnam.

I had no words to tell a three-year-old that death was a warm and safe conclusion to life, never wanting her to believe that if Heaven existed, it would be a better place. My three-month-old had no memory of her father, and I did not have the will to idolize a dead man, make him a saint, a more than human figure. I made do, acted bravely, created a new life, failed, and tried again. The rawness remained, festering. Not talking about it was one of the ways I numbed the pain. There was no ritual storytelling, no remembering.

In the four months since I learned American Legion Post 227 of Springfield Township, Delaware County, Pennsylvania, would honor him, a box of letters, cassette tapes, a wedding album, and the flag from his casket were fixtures on my dining room table. I read each word of the 104 letters he had written. I listened to both sides of each tape, my voice and his, and occasionally, the voice of our daughter Samantha, who was three-and-a-half that summer.

I remembered his devotion to us and our plans for the future. I opened the file folder of telegrams and citations: the pro forma letter from President Nixon and the condolence letter from Province Senior Advisor A.L. Kotzebue, who had sent him to the road that received small arms, mortar and rocket fire the day he died. So much had been conveniently forgotten, buried outside my heart. There were loud echoes of the question, "How did this happen?"

Daily, I refreshed the website promoting the event, scoured the collection of childhood pictures provided by his younger sister, and gazed at the face of my first love. I reread the carefully researched narrative on the American Legion's website, correcting mistaken information. I spoke frequently with Colonel Richard Debany, filling in details and educating him about life in the 1960s.

MEMORIAL DAY, 2019

I dress carefully for the day—a simple black sundress, not severe, but loose and comfortable. The air is warm but not as humid as I remembered the summers of 1970 and 71 when we lived in Baltimore and DC. Holly and I laid a wreath at the City War Memorial and joined a parade through the neighborhood streets. An honor guard from the American Legion Post 227 leads, followed by four former Army captains who were members of Advisory Team 88 with Bill. I walk with my daughters, Samantha and Abigail, with their spouses, and my two grandsons, Tristan, age twenty, and William, age ten. Meredith, my third daughter, has also come with us because we have been a family circle, entwined by love, sharing joys and heartaches, all affected by loss, for over forty years.

I am no longer twenty-four but seventy—our daughters are forty-seven and fifty. The Springfield High School band brings up the rear. It is a simple procession rather than a grand parade. Two more times, we pause, offer a prayer, and lay another wreath: Samantha and Abigail together in front of an Episcopal church; Tristan and William at the American

Legion; each time, tears flow while taps are played. Each time, we cringe at the gunfire of the military salutes.

William age 10, Tristan age 20

Well-meaning, patriotic supporters approach us from time to time, handing out small American flags. My hands remain empty as I mutter, "No, thanks. It's okay. I'm good." My grandson Tristan accepts the flags and then hands each one to the younger children along the route. At twenty, Tristan is the age of his grandfather when I met him. When he and his younger brother lay a wreath at the Springfield Township Building, I hear the gasps of recognition from Bill's classmates, who could see Bill in his big eyes, his cupid's bow grin, and the dimples on his cheeks.

"Thank you for his service and sacrifice," the woman says. She and her sister stand before me. They tell me their brother, who was a twin, also died in Vietnam.

I do my best to murmur thank you, wanting to say, "He was simply in the wrong place at the wrong time. Don't thank me. It wasn't supposed to happen. The war was nearly over, remember?" It is a day for a small town to honor a war hero. I am there to play my part. To demonstrate my bravery once again. I want to be thankful for this belated remembrance, yet I had struggled with myself to come.

After the parade, I listen carefully to the hour-long tribute, telling the story of a promising life cut short, watching pictures flash on a screen. Collectively, we thanked the community and presented our gifts to members of the American Legion who had orchestrated the details of the day.

I visit with Bill's high school classmates. I pose for pictures with members of Advisory Team 88—Mike Delaney, Ed Blankenhagen, Brian Valiton, and John Haseman surrounding me. I suggest they take me to lunch. I had read their names in Bill's letters. All of us are now in our early seventies. They had been young Army captains who survived the war, either stayed in the Army or came home, finished college, became lawyers and engineers, married, and raised families.

I ask them my unanswered question from that August day in 1972.

"He wasn't in combat. He went as an advisor. What happened? Why was he killed?"

These men had been advisors alongside Bill. They lived and worked at the team's headquarters in Ben Tre. As a Provincial advisor (S-2), Bill's work centered on intelligence reports. The others were district advisors and operations officers who were more likely to work directly with their Vietnamese counterparts, going everywhere together, placing themselves in danger, literally side by side. Each one was watching the war wind down, knowing it was coming to an end. These were the ones who gathered most evenings to go to movies or floor shows or drinks together and swap stories, tell jokes, and talk about their real lives back home. They were doing a job, meeting an obligation, fulfilling a duty, waiting to go home.

I carefully word my questions to them.

I said, "Obviously, Bill's death was a complete surprise. It wasn't supposed to happen. I have read the letter of explanation and the citation, but it has never been enough. Can you piece together the events of that day?"

In bits and pieces, I listened carefully to the normal crosstalk of conversation around a table. It wasn't the words that mattered but hearing sentiments of regret and personal loss.

Yes, I had read the comprehensive research of Colonel Rich Debany in preparation for the event. I knew by heart the specifics: Ham Long, the district north of Ben Tre, had been invaded in early summer, as Bill had written. Significant enemy forces remained there in August, cutting off the village of Tan Loi. Because of the urgency, Mr. Kotzebue's Provincial Senior Advisor needed to appoint an ad-hoc advisor to the threatened Ham Long District immediately— details I gained in Mr. Kotzebue's condolence letter.

As a temporary measure, Bill was sent to Tan Loi, chosen for his infantry background, training, and brief combat experience three years prior. As the Vietnamese troops moved along the elevated road, Bill was the senior advisor, the only advisor, and the sole American soldier. In an open area with rice paddies on both sides of the road, the enemy forces launched an ambush from the cover of trees. Bill was killed instantly.

On my right sat the senior officer of the group, Colonel John Haseman (retired), a district-level advisor in two districts in the Mekong Delta. He served two tours, spent two and a half years in Vietnam, and retired after thirty years on active duty, eighteen years in Southeast Asia. The advisory team in Ham Long, where Colonel Haseman had been the Deputy District Senior Advisor, was closed in May because of the perception that the security situation in Ham Long was improving. Beginning in July, Haseman was filling in as Deputy District Senior Advisor in Mo Cay, an adjacent province, while the assigned advisor was recovering from wounds to his legs. If he had not been in Mo Cay, Haseman could have been on that road on August 11.

On my left sat Captain Mike Delaney, a MAT (Mobile Assistance and Training Team) advisor whose tour of Vietnam coincided with Bill's. Both he and John were the ones who often accompanied Vietnamese counterparts on field operations, while Bill remained in the TOC (Tactical Operations Center), analyzing information and preparing briefing charts. If Captain Delaney had not been in Saigon on August 11, he was the other logical choice to be on that road to Tan Loi.

I was wrong in thinking I was seeking answers. I gained what I had needed all along. They, too, were shocked and devastated by Bill's death. He was the only American killed while they served on Team 88. They were the ones who mourned with me, who carried hard memories, whose lives were also altered.

Mike Delaney shared his memory of hearing Bill's baritone voice singing "Old Man River" every morning in the shower. I smiled. Our memories matched. He knew him well.

"I came to Vietnam through ROTC and a college degree," Mike said. "Bill was my equal: conscientious, smart, curious, and gregarious. He had a wickedly biting sense of humor. One of the hardest things I've ever done was deliver Bill's belongings to Saigon to send to you. I've always regretted that I never managed to contact you..."

"That's okay," I reassured him, "it was long ago. We are here now. I can only thank you for being his friend."

Sitting directly across from me was Brian Valiton, the Province Engineer Advisor. Brian spent most of his time off the compound, focused on engineering projects. He was not close to Bill but bunked three feet away with only thin plywood walls between them. Brian said little during lunch, but I knew he had made contact with Samantha over the years.

Brian said, "Bill quietly replayed the latest cassette almost every night when he thought everyone else was asleep. I never listened to the words. But I heard Samantha's voice, and I am glad I was able to meet her. I know your voice comforted Bill, cheered him up, and gave him a reason to live. When a new tape arrived, he was in great spirits the next day. Bill respected you and loved you very much."

Yes, my heart said, I knew that to be true. He maintained his integrity, was devoted and protective of me and our children, loved us, and wanted to come home to us. He had their support and friendship until the end of his life and beyond.

And so, at the end of a helpful lunch conversation, hearing that Bill was called in as a replacement to an unfamiliar village under siege to advise the District Chief and that he was the only American casualty, the tallest one with the radio, the easiest target, confirmed that all these years I had been right. What I said to the military envoys who came to break my heart on August 13, 1972—"It wasn't supposed to happen. He was only an advisor"—turned out to be true. My question to his compatriots was, "Why did it happen?"

As a chorus, they agreed, "It shouldn't have."

Their stories supplement mine. For all of us, the flame of memory remains.

When lunch was over, I joined my family at the home of one of Holly's school friends, whose parents still lived in the neighborhood where they had grown up. The current owners of 46 Longview Drive, Bill's childhood home, where I had lived for five months, had invited us to visit the house. I went with Abigail. The home had been remodeled, the garage was converted to an office, and Jayne's office and the sunroom were expanded into a family room. I still recognized it from the street, and my breath caught in my throat when I walked in. The front door opened into the living room. I entered and glanced up the stairs. Polite conversation and hospitality were offered before I could ask if I could see the room at the top of the stairs. That space had been Bill's room, which became our room in November 1968, with an attached room barely large enough to hold a crib.

As I reached the top of the stairs, my heart raced. I was twenty years old again. Tears of loss sprang to my eyes. Of course earlier, there had been some unbidden tears caused by the sound of taps, the laying of a wreath, and listening to the catch in Samantha's voice as she read her father's name. I was haunted as I walked into the bedroom we had shared in Bill's childhood home, seeing another crib for other grandchildren. It was where I had first laid Samantha when we brought her home from the hospital.

Back out on the street, I turned to Abigail. "I am so sorry I have never spoken of your father. I always felt that if I kept moving, living, started over, and never revisited the past, I could be happy," I said. "I never wanted you to feel my pain."

"It's okay, Mom," was all she said.

I took her in my arms, remembering the comfort it gave me, embracing the baby whose heartbeat carried her father's lifeblood. Indeed, this Memorial Day was long and filled with pain and healing. It was the end (or was it merely a resting place?) of a long journey of grief.

Through conversations, questions, and confrontation, sorrow twisted my heart and memories seared my soul. Having my daughters and grandsons with me was a pure gift, and I am grateful that they accompanied me. I became aware that the event was also essential for their understanding and healing. As well as his four comrades who served with him, we each carried memories, questions, regrets, and sorrow.

On Tuesday, May 28, I woke up at 4 a.m. and let the tears flow. Emotionally exhausted, I saw the somber faces of two long-forgotten soldiers who brought the news of my husband's death to me on an August

Day; I recalled the shared grief of four no-longer-young Army captains who were stunned by the loss of a friend and came from Colorado, Massachusetts, and Kansas to honor him; I gave thanks for my daughters and grandsons who walk with me through life and give me meaning. I suddenly realized my grief had been shared by others all along. I gave up the "if onlies" of my long life and began to celebrate "what is." I have never been alone.

I allowed the truth to wash over me. Forty-seven years have passed. Yes, it was a great sacrifice, a significant loss. Still, I am not a believer in the idea that war, any involvement in death and destruction in another sovereign nation, is in the best interests of this nation or those asked to make the ultimate sacrifice.

CHAPTER 29
TRUTH RECLAIMED

I SAID GOODBYE to my family at the airport. I took time to drive past Marcus Hook, where Bill had embarked on his ship with the Merchant Marine, sailing across the Atlantic while I was traveling throughout Europe. I drove myself to Longwood Gardens, where Bill loved to take closeup pictures of the flowers, a place we brought Samantha in a stroller, a place for me to sit alone, remember, sort through the weekend, and write.

On Wednesday, May 29, I left Philadelphia to fly to visit Merry-K, my childhood friend, still living in Durham after a career as a professor at the University of North Carolina. We seldom saw each other over the last fifty years. Our annual exchange of birthday cards and stories that filtered down through our mothers and school acquaintances kept us informed. We reconnected at our fifty-year high school reunion. She offered me space to decompress from the intensity of the weekend.

When we drove from the airport, I was shocked to hear her tell me she remembered little about my wedding. She said, "I was dismayed. I didn't understand why a smart girl like you would leave school to get married."

I was startled, realizing how carefully I had guarded my pregnancy secret from even my oldest friend. I laughed at the notion that I was anything less than transparent once upon a time. The floodgates opened to tell her about the weekend. We spent the next five days reminiscing our childhoods and time together in Washington, DC, recalling our nights of dinners and playing bridge, and remembering two-year-old Samantha sensing Merry-K's uneasiness around her, asking, "Do I make you nervous?"

On long car rides, late nights, and through constant conversation, we celebrated being rooted in a small town's safety, security, acceptance, and nurturance. I rehearsed the stories prompted by the Memorial event, lamenting how Bill's request to "Be brave" had sometimes stifled my need to grieve.

Merry-K named a better watchword for me. "Resiliency."

Yes, my heart said, "I am Resilient. We are Resilient"

Me and my daughters
July 25, 2018
Samantha, Abigail, Bonnie, Meredith

RETROSPECTION

I N TWENTY-SEVEN YEARS of professional ministry, I understood my vocation was to teach and illuminate universal truths and provide spiritual insights to transform lives. I have worked for equity, justice, the protection of the vulnerable, and inclusion of the marginalized. I have listened and helped others find their answers. I always hung two prints in my office as reminders of the pain of my losses, as inspiration and comfort, signposts of resiliency.

One, an artwork poster, illustrates every symbol used to identify prisoners of the Germans during the Holocaust, each symbol labeling them for extermination. Beneath the six-pointed stars and variety of colored triangles is this quotation from the Talmud:

"Whoever destroys a single life destroys the entire world.

Whoever saves a single life saves the world entire."

Nearby hung a calligraphy print of these words, attributed to Thomas Gray:

"If I should die and leave you here alone,

be not like others sore undone, who keep

long vigils by the silent dust and weep.

For my sake, turn again to life and smile

nerving thy heart and trembling hand

to do something to comfort weaker hearts than thine.

Complete these dear unfinished tasks of mine

and I, perchance, may therein comfort you."

As I have been writing these stories intended for my daughters and my grandsons, I realized a desire to reach a larger audience. I heard my preaching voice. I spent years exploring Biblical narrative, making ancient myths and gospel stories relevant to current reality. I never sought sympathy because of the course my life has taken, although frequently I found myself preaching against violence, illustrating with pieces of my life experience.

The imprint of Viktor Frankl's writings I encountered in college influenced my philosophy then and still sustains me. In *Man's Search for Meaning*, he wrote about being asked to address his fellow prisoners in Auschwitz. On a night surrounded by emaciated men, in their cold beds, Frankl speaks of hope "Human life, under any circumstances, never ceases to have meaning."

I have never found meaning in the death of my twenty-four-year-old husband, father of two, once alive with hope and endless possibility. Year after year I was aware of the date of August 11. I annually wrote pages in my journals about how our dreams may have been lived out, how much our daughters would have loved their father, and how I may have known deeper happiness. I have always found comfort in having been loved by him—"a bloom of humanity plucked from life—unique and irreplaceable" (words from his memorial service, August 24, 1972). Strengthened by the ideal of stoic bravery, buoyed by an interior resiliency, I have made meaning of my life.

My fifty years of silent grieving have been lessened by knowing the friendship of his fellow soldiers and the honor he was given by the American Legion of Springfield, Pennsylvania, on Memorial Day, 2019. In Colonel Rich Debany's words:

Bill's life was much more than his final act. He was a loving husband, a caring and proud father of two daughters, a thoughtful son, and a brother to three siblings. He was a great friend to many in his hometown of Springfield, Pennsylvania, and to his fellow soldiers with whom he served. Dedicated, he was an officer who professionally did his duty with honor and distinction despite private misgivings about our nation's involvement in Vietnam.

His legacy isn't defined, nor is his story punctuated by his heroic and selfless death... he defined his legacy through his meaningful life, the people he influenced, and the positive differences he made. His legacy is the rich and fruitful lives of his wife Bonnie and their daughters Samantha and Abigail. It's the lives and memories of his friends from Springfield and Team 88, U.S. Military Assistance Command Vietnam.

EPILOGUE
LETTERS NEVER SENT

13 August 1972

Letter #35 (since July 10)

Dearest Darling,

I know this is foolish and probably a sign of my mental instability at this moment, but after all, I've already written you two letters since you died, so why not one more?

It's been a strange and horrible day. Everything was fine, and routine, and tranquil. Samantha and Stephanie were gone to Sunday School for an hour with Grandma, and Jim and Vicki cleaned up the front room of the mess of toys while I did up the breakfast dishes. Ruth came down about 11:30 to make arrangements for paying for the truck tomorrow. Clem and Dad were working on some parts for the truck. All was normal; life was fine.

Sometime around 12:30, things began to fall apart. An Army-type official car entered the yard, two men (an officer and an NCO) got out, asked Dad if I was there, then came into the house. I knew the instant I saw them; I heard the words before they came out of the Colonel's mouth (only a Light Colonel, though). "The Secretary of the Army regrets to inform you...etc." I shut my eyes to blot it all out—but nothing could or can make it go away. It's true, I know it, but how can it be? To never be held by you again—to have you never see and love your girls. Oh, Bill, I won't believe it, but know I must.

Time has gone by very slowly today, and we've had a steady stream of love and concern shown. First, Ruth and Clem, and then the phone calls to and from various people. Vern and Sally came by with dinner; Rhoda and Becky (whom I don't remember you ever meeting, but maybe so); her three kids, Gene and Marian Moos; and Bill and Marilyn. Lois and Ellen both called, as did Aunt Esther from Seattle. And finally, at 9 o'clock, when your Mom found out, she called too. Mom called Parm so that she could let Merry-K know. We called the Red Cross to let Sue and Dick know, and I called Gordie and Arlene to tell them—and they were extremely upset. Everyone has been great, but the number of tears shed has been exhausting. I don't know how you would feel about this show of sympathy, but I keep imagining you being amused, and I occasionally detach myself to be with you mentally.

Samantha doesn't know or even suspect anything is wrong. She's so smart and curious that she would ask unanswerable questions, which wouldn't mean much to her. Right now, she's just used to you being gone, and this winter, when she begins expecting you home again, I'll have to explain it to her completely. How, I don't know. And what her memories of you will be, it will be a long time before I know. I'm sure she will remember you, though. You loved her dearly, and she adored you. Abigail will never have any idea of you at all, but when she is old enough, I'll tell her all about her daddy. My love for you continues on and on. My memories are going to last forever. Never in my life will there be anyone who means so much or who can be so dear or so kind or, well, at times, naughty or fun-loving. No one to make me so mad or so happy or so completely fulfilled.

I know our lives will go on, though. I'm going to try to do the wisest thing with all the insurance and go back to school and finish and somehow support our girls. The idea of leaving them bothers me a great deal, and maybe it will take me five years to complete, so Abigail will be ready to go to school soon. Time will tell, and there's lots to iron out. Lots of pieces to pick up, but for you, I'm trying to be strong, and for our girls, too.

I'll always love you best and cherish every memory and thought of you, but everything seems hollow and empty and shattered now. How long before the sun shines again? How long.

Always yours,

Bonnie

12 May 2001

A Visit to the Vietnam War Memorial

Dearest Darling,

I thought I was ready to make this pilgrimage, seeking healing at the deepest level. I was afraid, not wanting to go to the place where it hurt, not wanting to remember the loss and death of my dreams; I didn't want to cry and not stop; I didn't want to start feeling angry or afraid again. After spending years stifling my tears, my grief, and sadness so that maybe you would "come" back to me as my father reported experiencing your presence in a wheat field after visiting your gravesite at Willamette National Cemetery in December and acknowledging that death is final; after recognizing how early I was programmed to be dutiful, brave, and strong, obedient and devoted; after declaring I am not willing to let anyone make decisions about my life, I came hoping to accept the destruction of my innocence, my loss, my despair, my trust-filled center at 24. It has been twenty-nine years of loss and longing.

I carry the picture of you, taken on the day of your death. I captured it from the internet and printed it at home before I left to come for a conference in Washington, DC. Last night, I wrote on the backside. Partly a letter. Partly a story of who you were, hoping someone would pick it up, file it away, preserve it, and read it. National Park Service Rangers come daily to remove items left at the base of The Wall. Will it be archived, analyzed, studied, or published one day?

I approach Panel 01W, looking for Line 62. I know exactly where to find the panel and your name on the wall. The panels meet, with the beginning date of 1959, honoring the first American casualties moving from the center to the east—the final panel, dated 1975, marking the last. Yours is one of over 58,200 names. There are too many names written so small. Too many children around me are unaware of what it all means. There are people beside me, crying, accepting their losses. There are individuals with no connection to one another brought together by death. Questions are being asked, and doors to understanding are being opened.

I am surprised by the tributes—wreaths from schools, classmates, and children; single flowers; notes on lined paper; drawings by children who could never have known the uncle or grandfather. I want to point, tell people, have them look, remember, not at all, but just one name. I linger, never wanting to move, never wanting to repeat goodbye. Your name is engraved in granite; there is no denial. Yes, you are there. Yes, you are dead. Yes, I miss you. Yes, a nation knows. The tears come, primarily for you, some for myself, some for the continuous waste of human life in war, knowing grief is multiplied worldwide, throughout history, every day.

I ask for paper and a crayon, take a rubbing of your engraved name with me, snap some pictures, and walk away, clinging to memories.

I still love you best and will love you always,

Bonnie

13 August 2023

Dearest Darling,

It has been a long time since I have written, or as we used to say, "Only forever." It has been fifty-one years to the day since I learned that you had been killed in Vietnam. This is also a Sunday. I just finished washing the breakfast dishes, and as it often does with physical activity, my mind can roam, come up with new awarenesses, and have bright ideas.

Today, at 1230 hours, I remembered a moment in the kitchen on the farm. We were newly married. You were holding me in your arms. I leaned back, looked directly into your eyes, and said, "I have three questions."

With a deadpan look, you replied, "I have three answers." You paused briefly, then with a twinkle in your eye, you said, "Yes, No, and Maybe."

Through my laughter, I said, "Now I have to ask the questions in the right order."

I still hear myself announcing, "I have a question."

In the writing of these pages, in light of this anniversary, standing at the kitchen sink, I realize some questions will never be resolved, yet others have been answered.

All this time, I have wanted to know how you could die. I asked those who brought me the news, and they handed me a telegram. I received Official letters that offered sympathy and gave me clues, but I didn't know how to follow the clues. I told myself it didn't matter but imagined scenarios, watched too many documentaries and movies, and read others' accounts of combat.

Did you know the blast of a B-40 rocket-propelled grenade killed you? Was there a moment of awareness that you would die? Did you suffer? I pray the answer to each of those questions is "No."

Most recently, I learned it took eighteen hours to recover your lifeless body from where it lay. The unit you were with retreated after the ambush, leaving you and their other casualties behind, and returned to district headquarters. It took the ARVN 18th Regiment reaction force a while to assemble, figure out how to sweep the ambush site without being ambushed, and then conduct the operation.

There is a poetic irony in the fact a Memorial Service was being held by your team members in the Team 88 Compound in Ben Tre at the approximate time military representatives arrived at a small farm near Edwall, Washington, to let me know you were dead.

Your remains were moved to the U.S. Mortuary at Tan Son Nhut Air Base in Saigon. Mike Delany and Dave Godby, your evening drinking buddies, sadly transported your personal effects to Saigon. Sadness for all of those who knew you overwhelms me. (Maybe you already know these things from that thin place between this world and you.)

How I wish you knew our daughters and they knew you. They are strong, bright, curious, intelligent, kind, and fun-loving. They acquired your love of reading and appreciation for music, art, and the theatre. They are both educators with an informed distrust of the military-industrial complex and an ingrained distaste for gun violence. Abigail has given us two handsome grandsons, Tristan and William. You would be proud.

What I want to tell you now is my declaration: "Never in my life will there be anyone who means so much or can be so dear or so kind... or fun-loving. No one to make me so mad or so happy or so completely fulfilled... I'll always love you best." All of this came true.

Although I remarried quickly in December 1973, my choice of another Vietnam veteran, whom I hoped would understand my grief without ever needing to talk about it, was ill-advised. We had fun, and he made me laugh as he played with the girls. Laughter surprised us. At four and a half, Samantha had forgotten about hearing laughter. Abigail, in her first year, maybe heard none but her own. I gave birth to another daughter, my blue-eyed blond, named Meredith Leah. As Samantha's middle name honors your mother, Merry's name honors mine. She is a person of wisdom, creativity, and kindness who has given me three beautiful grandchildren.

The color TV in our family room on January 20, 1981, announced the release of fifty-two American diplomats and citizens after 444 days as hostages in Tehran, Iran. They received a heroes' welcome. Not one died.

Not one had risked their life in combat. Unlike you, whose body was quietly flown to California, cremated, and your remains interred at Willamette National Cemetery, with only family to witness, a small town to support us. You were mentioned in a brief news story on TV. No parade. Not for the living or the dead. My grief returned. Full force. There was no understanding of my anger, no room within the chasm between us. My second husband was an impractical dreamer and schemer, unable to manage money and soon declaring bankruptcy. We divorced.

In 1983, I tried again, marrying a man who had been an Army Captain stationed in West Germany for four years. He had a college degree and a Master of Divinity degree. We shared a commitment to the work of ministry and a passion for justice. Like you, he loved to read, and we enjoyed movies, theatre, and music. He listened well and empathized, accepting my pain and honoring my loss, although I was still unwilling to talk about it.

My patience wore thin because of his inability to manage money; I always felt financially insecure. My heart was burdened with a desire for him to love me enough to care for his health. I was frightened, expecting his death from obesity, heart or kidney failure, for most of his final decade. We were together for thirty-four years until he died of a heart attack. Again, I became a widow.

I realize I was attempting to hang on to the dream, expecting to find in former soldiers the integrity, ambition, sense of responsibility, and devotion I experienced with you. Neither of them stood a chance to make me completely happy. Or restore my faith in a frightening world. Or help me feel safe.

Widowed again, I began to write the story of our life together. At sixty-nine, I had never lived alone. From the farm to school, then marriage and family, my entire life was lived with other people—someone to talk to, someone to care for, someone to laugh with. Being brave and maintaining cynicism throughout the Memorial Day event in Springfield, my defenses broke, and I reclaimed my softness and vulnerability, the ability to trust and to live again. I embraced resiliency, gave thanks that, surprisingly, my Social Security benefits are the highest rate, based on your projected income, and discovered the power in my financial security. And yes, I married again. Now, I have found a sense of safety and the devotion of an educated, kind, considerate, well-read man who has the desire to care for me and the means to provide for me. He is willing to listen to me every morning and night. I fill him in on the details of my day, how I feel, and what I've been thinking. I live with a restored sense of honor and respect.

Although my luggage is no longer part of a matched set and is burgundy in color, over the years, I have traveled to Hawaii as often as possible, making up for the time we never had on the island of Maui. I missed you being with me as I stood on a beach in Maui for Abigail's wedding.

On every trip across the country, I remember our travels together. I visited places you had been to and only told me about—the Winchester Mystery House in Monterey, California; Stone Mountain in Georgia—always thinking there might be something of your essence, a sharing of your experience there—a thin place.

Heaven and earth, the Celtic saying goes, are only three feet apart, but in thin places, that distance is even shorter.

In September 1973, when Father was planting the following year's crop, he came into the house, shaking his head. He said, "I saw Bill in the field while plowing today."

"What are you saying? What do you mean?" Mother asked.

She knew he wasn't a foolish person. He wasn't a drinker. She didn't consider him religious or superstitious.

"I saw Bill. He walked toward me. I heard him say,'Would you tell Bonnie I'm all right? When I've tried to tell her myself, she always starts crying. Let her know I'm okay.'"

Hearing this comforted me. I knew I needed to stop crying every night. I thought it might help. If I visited places we had been together, I would feel your presence and remember you more clearly. Perhaps if I saw the places you had been without me, maybe if I were still.

It never worked. Memories faded. I have spent the first two weeks of each August at the Bead Lake cabin, where we watched stars and satellites cross the sky during your final visit home, dreaming of our future. I journaled annually, always wondering how our fate would have evolved. I thought about the life I had made for myself and our daughters, analyzing all my mistakes along the way and knowing that you would be proud of who your children had become and how much you would love your grandsons.

I searched for you for another forty years until Abigail took her two sons to Willamette National Cemetery.

We each carried a red rose. William, age five, kneeled, ran his finger over the letters, and said, "His name is my name too."

In his cupid's bow grin, the spark in his hazel eyes, the cowlick on his forehead, heaven and earth met again.

In my love for you, they have never been separated.

Our love story, your legacy, continues.

My memories will last forever.

Your Kitten, always,

Bonnie